KILLING IT

Camas Davis is a former editor and writer for magazines including *Saveur* and *National Geographic Adventure*. In 2009, she travelled to southwest France to study whole animal butchery and charcuterie and subsequently founded the Portland Meat Collective, a transparent, hands-on meat school that has become a resource for education and reform. In 2014, Camas launched the Good Meat Project, a nonprofit dedicated to inspiring responsible meat production and consumption through experiential education across the USA.

CAMAS DAVIS

KILLING IT

A Memoir of Life, Love, Death and Dinner

PICADOR

First published 2018 by Penguin Press, an imprint of Penguin Random House LLC, New York

First published in the UK 2018 by Picador

This edition first published 2019 by Picador
an imprint of Pan Macmillan
20 New Wharf Road, London N1 9RR
Associated companies throughout the world
www.panmacmillan.com

ISBN 978-1-5098-1102-1

Portions of this book, some in different form, appeared as part of "Girl with Knife" in Double Bind:
Women on Ambition, edited by Robin Romm (Liveright, 2017), "Human Principles" in Ecotone,
and "Know Where Your Food Is From" in MIX Magazine. "Run Rabbit. No, Really, Run!" a
feature in an episode of This American Life was adapted from "The Messy Middle," which appeared
in Oregon Humanities. A quote from Charles Eisenstein's "The Ethics of Eating Meat" which
first appeared in Wise Traditions in Food, Farming and the Healing Arts, the quarterly magazine
of the Weston A. Price Foundation, Summer 2002 appears here in the epigraph. A quote from
John Berger's Pig Earth (Bloomsbury, 1999) appears here in the epigraph.

1 3 5 7 9 8 6 4 2

A CIP catalogue record for this book is available from the British Library.

Typeset in Fournier MT Std
Printed and bound by CPI Group (UK) Ltd, Croydon, CR0 4YY

Visit **www.picador.com** to read more about all our books
and to buy them. You will also find features, author interviews and
news of any author events, and you can sign up for e-newsletters
so that you're always first to hear about our new releases.

FOR DJUNA,

AT THE VERY BEGINNING

OF *ALL THIS*

You road I enter upon and look around, I believe you are not all that is here,
I believe that much unseen is also here.

WALT WHITMAN, "Song of the Open Road"

Despite our machinations, we are ultimately unsuccessful at avoiding
pain, loss and death. For animals, plants, and humans alike, there is
more to life than not dying.

CHARLES EISENSTEIN, "The Ethics of Eating Meat"

"Life," as the Russian proverb says, "is not a walk across an open field."
Experience is indivisible and continuous, at least within a single lifetime
and perhaps over many lifetimes. I never have the impression that my
experience is entirely my own, and it often seems to me that it preceded
me. In any case experience folds upon itself, refers backwards and for-
wards to itself through the referents of hope and fear; and, by the use of
metaphor, which is at the origin of language, it is continually comparing
like with unlike, what is small with what is large, what is near with what
is distant. And so the act of approaching a given moment of experience
involves both scrutiny (closeness) and the capacity to connect (distance).

JOHN BERGER, *Pig Earth*

PART 1

PART 1

ONE

She was big. I didn't even know pigs could get that big. And although I could see for myself the astounding girth of her, I had no
way of wrapping my head around the sheer physical reality of such a
weight, until the mechanical hand that had lifted the dead old sow up
out of the concrete bath of scalding water accidentally dropped her,
from five feet high, onto the hard, cold concrete floor. It wasn't a thud,
exactly. It was more of a ripple. A reverberating ripple of fat and skin
and bone. Her heart was still in there, too, though no longer beating. So
were her kidneys, her spleen, her lungs, her gallbladder, her small and
large intestines, and everything else that had once made her alive but
now made her a very heavy carcass on the floor. Without flinching,
several men in dark-blue coveralls scrambled over to her body and attempted to push her back into the mechanical hand. As if this were actually possible. It most certainly was not.

"She's three hundred kilos. Too big," Marc Chapolard, who, along
with his three brothers, had agreed to mentor me in the French ways of
knife and bone, told Kate, my American translator and host, who told
me. I did the math. Almost seven hundred pounds.

Just minutes before this pig carcass had accidentally fallen to the
floor, a tall, thin, older gentleman, also in coveralls, had escorted the

live sow through a wooden chute. As I watched her slowly make her way—she had so much *body* to move—Marc explained that she was done having babies and was going to be turned into sausage.

Another man then secured what looked like a set of headphones onto the sow's head.

"Are they going to play her music?" I asked, in all seriousness. No one answered me.

Then, with the meaty part of his palm, the man pressed a big red button on the wall, which sent a quick electrical current coursing through the headphones and into the skin and the subcutaneous fat and the bone and then the brain of the seven-hundred-pound mama, who dropped to the floor and began convulsing.

This felt to me like a private moment that I shouldn't be allowed to witness, these last shudders of life, this battle waged by her body's nervous system, but I did not turn away. I was here to learn, after all. I was here to do something hard and real. I wasn't interested in wasting any more time pretending that what was unfolding in front of me wasn't really happening. I'd spent the past ten years of my life doing that, and I was determined never to go down that road again. But words like *hard* and *easy* didn't seem to apply here. To the people working in this abattoir, to my French mentors raising pigs and butchering them back at the farm, this scene—would they even call it a scene?—was, quite simply, work. Life, death, work. So I kept watching.

"That makes her senseless to pain," Marc told Kate, who told me. This was the most humane way to kill a pig, he said. Sever all communication between the brain and the nervous system as quickly and painlessly as possible.

"She's not dead yet?" I asked.

"No," Marc told Kate, who told me. "She's unconscious. She won't be dead until they bleed her."

When I looked back, another man had hoisted the pig up by way of a chain wrapped around her back leg. He stuck a long knife into the space between the sow's throat and where I imagined her heart to be.

"She can't feel this," Marc said in French, Kate translating. "If we didn't stun her first, she would feel it."

A woman held a plastic container underneath the sow to collect the blood. We all watched in silence as she rhythmically stirred the blood to keep it from coagulating. Time slowed to that rhythm. My brain slowed to that rhythm. My breath slowed to that rhythm. It was nowhere near a lullaby, but something in this woman's focus, in her unwavering attention to this one small detail, something in the stillness of her face, almost soothed me. Almost.

But as I looked from the woman's face to the sow's now lifeless visage, I flashed upon an utterly inappropriate metaphor, one that linked the dead seven-hundred-pound sow to the very much alive one-hundred-thirty-pound me. This sow, I thought, had spent her entire adult life doing her job well. Her job was to make babies. Lots of them. Babies that could then be turned into food for our tables. And then, one day, someone had deemed her no longer useful. And just like that, her job had ended. Hence the headphones and the electric shock and the blood. At least she could still be used for sausage.

I'd spent my entire adult life doing my job well, too—although it didn't have anything to do with making babies for the dinner table. And then, one day, just a few short months ago, someone had deemed me no longer useful and my job had also ended. It had all happened so fast. Given the way things had gone in my life recently, I felt like I was the one who had just been dropped from a five-foot height onto a hard, cold concrete floor. But of course, I hadn't. Of course, I still had blood pulsing through me, and this sow didn't have any. She couldn't feel anything anymore. But I could. And I knew that, in my attempt to relate my narrative to hers, I was only distancing myself from what was actually happening in front of me: the transformation of life into death into food.

Marc nudged me over to the bucket of blood.

"This week," Marc said, "you'll help us make *boudin noir*."

Blood sausage. I had eaten it several times as a food editor and restaurant reviewer back in the States, but I'd never actually seen anyone prepare it, let alone made it myself.

After less than a minute, the sow's bloodletting ceased. She was dead. There are so many other ways to say it or think it, all loaded: She had no more life to give. She had passed. She gave her life. Someone took her life from her. They killed her. They slaughtered her. They harvested her. They murdered her. Her body had the potential to become heavy with the weight of too much meaning. Her body had the potential to become heavy with the weight of the absence of meaning. But here, in this moment, she was, quite simply, dead.

The men pushed her heavy carcass into a deep concrete pool of steaming hot water, to loosen her hair follicles, Kate explained. Two men, with the help of long paddles, rolled her carcass gently through the bath, shepherding her toward her next fate, that mechanical hand. As the mechanical hand scooped her body into the air, Marc explained, it would twirl her body around, as if on a rotating spit, while hot flames shot out and the rubber fingers of the mechanical hand massaged the hair right out of her skin. But before any of that could happen, there was that drop. The seven-hundred-pound ripple. The men scurrying to try and push her body back in.

"Why don't they just skin her?" I asked, thinking of the deer I'd watched my dad skin as a child.

"Because the skin is food," Kate answered. "We don't waste food here." Earlier in the week, I'd discovered little rolls of pigskin tied with red-and-white butcher's twine in Kate's piggery, or pantry. She stored them in tall, sealed jars filled with rendered pig fat to keep oxygen from spoiling the skin. "For cassoulet," she'd told me. In Gascony, an oft overlooked corner of southwest France and the place I would call home for the summer, cooks line their *cassoles*—the upside-down-bell-shaped ceramic pots that cassoulet is traditionally cooked in—with pigskin before they place Tarbais beans, duck confit, and pork sausage into the pot. The pot is then placed in the oven for hours, until everything becomes tender, oozing with the flavors of meat and fat and beans. The skin not only keeps everything from burning or sticking to the pot; it provides the collagen and fat that's responsible for the rich flavors of the dish.

One of the men in the abattoir pulled out a small blowtorch and a couple of knives so that they could scrape and burn her hair off by hand. And in this next moment, a moment wedged between the finality of her death and what was to come next, the further transformation of her body into food, the real, visceral scent of scalded pigskin mingled with the imagined scent of cassoulet in my brain. The gap between the two closed just a little, but not completely. Even in this abattoir, facing the stark exchange of death for dinner, that gap refused to go away. There remained, somewhere between the genuine article of this sow's life and death—somewhere between her death and dinner—a black hole, an undoing.

Marc ushered me to a neighboring room and pointed to buckets of hearts. Buckets of spleens. Buckets of lungs. Everything could be used for food, he told me. In another room, lit by a single lightbulb hanging from the ceiling, a lone, stout man, with tufts of gray hair spilling out of the collar of his white butcher's coat, cleaned pig intestines over a porcelain sink. The smell of bleach and vinegar wasn't quite enough to cover up the organ funk that permeated the room.

"These will be used for stuffing sausages," Kate told me.

This was for the most part being done the old-fashioned way, on an incredibly small scale, with just a little help from mechanical ingenuity. Before coming to France, I'd seen a statistic that 99 percent of all the animals raised and slaughtered in the United States were factory-farmed. I assumed that statistics for the rest of the industrialized world, including France, weren't too far behind. The animals in this tiny, cooperatively owned abattoir, however, represented the other 1 percent. The floors were concrete. The walls were limestone. The doors were open. The people who owned the slaughterhouse were also the farmers who raised the animals and, in some cases, the butchers who turned the carcasses into roasts and hams and sausage and bacon, or *ventrèche*, as the Chapolards called it. There was little division of labor here. Farmers like the Chapolards shepherded their animals all the way from seed to sausage. This lone man standing underneath the lightbulb opened the ends of the intestines with his two sausage-size thumbs and held the openings under the faucet to clean them. Then he massaged them with salt and vinegar until they were ready to be stuffed with ground meat and fat. That's how they did it in Gascony. It didn't seem to matter that it was a lot of work. An animal's life had been taken, and the animal deserved this kind of attention to detail.

I could feel Marc and Kate looking at me, trying to gauge my reaction. Somewhere just beyond my conscious thought—*So this is how it's done*—that black hole swirled. And it was into that black hole that my limited understanding of life and death and all the emotions and thoughts one is told to have or not have about such things disappeared.

"What does she think?" Marc asked Kate, who asked me.

I furrowed my brow. *"Je ne sais pas,"* I said. I closed my mouth, then opened it again as though I had something more to say, but I didn't. The distinct flavor of wet barnyard and blood hit my tongue. *Je ne sais pas.* I don't know. It was one of the few things I *did* know how to say in French, and it was true. I didn't know. I'd just spent the past ten years of my life making sure I knew everything about everything. I was paid to know. If I didn't know, I was paid to find out. And then I was paid to write what I knew and come up with a clever headline so that everyone else would know, too.

But today, I didn't know anything. I wasn't sure what I felt. It was as if, because no one was paying me to write it all down, I had no words to describe it even to myself. And without words, swimming around in that black hole between death and dinner felt deep and vast and lonely.

TWO

One week before I stepped into my first abattoir in France—before I stepped into my first abattoir anywhere—I left my home in Portland, Oregon, took a long plane ride across the country to New York City, and then boarded another plane to Toulouse, in southwest France. On the plane to Toulouse, my seatmate, an attractive Frenchman with smooth olive skin, well-coiffed hair, and a neatly folded silk handkerchief in his breast pocket, asked me, in perfect English, what I would be doing in France.

"I was a magazine editor for ten years," I told him. "But I lost my job. And now I'm going to study butchery in Gascony."

My seatmate's eyes widened. I sensed that my answer had interrupted his perfect sense of order in the world. His silk handkerchief nearly wilted in disappointment.

Since this already seemed like a big enough pill for the man to swallow, I refrained from mentioning that, in addition to recently having abandoned my ten-year career as a magazine editor, I'd also ended a ten-year relationship with the man I thought I would marry, that I'd promptly moved in with another man with whom I thought I was in love, and that, aside from my weekly unemployment checks, I was completely broke, save for an unused credit card I'd found in the back of my filing cabinet, which, against the better judgment I once possessed, had paid for my seat next to him on this plane.

I was a magazine editor. I lost my job. I ran away to France to become a butcher.

I liked the sound of it. But the man with the silk handkerchief did not.

He looked me up and down and said, rather exasperated, "But you are such a beautiful woman. Why would you do that?"

I suppose a different kind of woman might have been flattered, but I was not.

At least he'd bothered to ask me *why*. Back home, most people I told this to pretended they hadn't even heard me. Or, sometimes, they'd ask whether I was joking.

There were exceptions, of course. There was Will, the furniture maker I'd jumped into bed with shortly after ending my ten-year relationship with Tom. Will, whose half-finished house I'd recently moved into after losing my job as the food editor and managing editor of a scrappy city magazine in Portland, claimed he understood. Even if he was disappointed that I had to leave the country to pursue my interest. Even if he sensed that my decision signaled a deeper unrest.

My oldest friends, who, over the past twenty years, had grown used to watching me make unorthodox decisions—*I'm going to Nicaragua to organize female farmworkers; I'm going to teach creative writing in a women's prison in Rhode Island; I'm quitting magazines to get a graduate degree in performance art at NYU; I'm going back to magazines as a food writer*— were rooting for me.

And my father, a lifelong hunter and fisherman, who handed me my first pocketknife and taught me how to gut a fish when I was nine, was excited for me to eventually share my new skills with him—that is, if he ever brought an elk or deer home again. Ever since he'd switched to bowhunting in my adolescent years, the pantry shelf normally reserved for his homemade venison jerky had grown bare.

On the other hand, my mother, a woman who ate and cooked meat on a regular basis but refused to touch it in its raw form—when my father wasn't around to transfer the chicken breasts from their Styrofoam package into the frying pan, she relied on gloves or tongs—strongly objected.

"I can't believe my baby is going to become a killer," she said more than once. "It sounds so dangerous."

And then: "When you come back, are you going to get a real job?" she'd ask. A *real* job. As in a woman, a pencil skirt, a laptop, a lifestyle magazine. Legitimate, sensible work for a modern, urban, thirty-something woman like me.

"I don't think so," I told her.

Most people, however, simply met my news with confusion, usually in the form of a blank stare. It was as if the very word itself—*butcher*—was so completely foreign to them that they did not deem it worth their energy to comprehend.

I WAS USED to people responding to me this way. I have an unusual first name, such that if, upon introduction, I tell you my name and you just nod your head and move on with the conversation, I know that you have no idea what my name is and probably don't care to know.

But if, like me, you are the sort of person who is interested in the stories behind the names of things and you ask, "What kind of name is that?" I'll tell you the long or the short of it. That it's a wildflower that grows prolifically in the Pacific Northwest, where I grew up. Or that it was a major staple of the region's American Indians, who dug up the roots and baked them in the ground like potatoes. Or I'll tell you it's a wild lily, *Camassia quamash,* and that there are two kinds—the purple, edible kind and the white, poisonous kind—and that they bloom for a very brief period in the spring.

Or, if you really seem to like a good story, I'll tell you that when my mom was pregnant with my twin brother and me, she was driving south on I-5 with my dad, out of Eugene, Oregon, where, a few months later, Zach and I would enter the world, and they passed a sign that said CAMAS SWALE. They thought Camas would make a nice name for a girl, except they didn't want to name me after what amounted to a drainage ditch that ran under I-5. It wasn't until they came upon a field of purple camas flowers in Idaho and someone identified the flowers for them that they decided to name me Camas, after the flower—the stem, the petals, the pistil and stamen, the root, the food, the poison.

"So I'm really named after a ditch," I'll tell you.

"That's not really true," my mother always reminds me. But I tell the story anyway.

If you want me to keep going, I might even tell you I actually rather like having been named after a natural or man-made depression whose sole purpose is to collect the world's runoff, filter it, and then spread it horizontally across the landscape. Concentration, filtration, diffusion. Editor. Butcher. There you have it.

Just as when I tell people my name, when I told people I was going to France to study butchery, I was attempting to tell them a story. What was it about this particular story, I wondered, that confounded so many of them?

OF COURSE, if I'd had a sense of humor back then, I would have understood how funny it sounded: *Girl breaks her own heart. Girl loses job as a magazine editor. Girl seeks a more authentic life. Girl starts killing animals.* But broken hearts and wrecked careers have a way of turning a girl rather serious, so it didn't feel all that funny to me when I waved goodbye to everyone and flew to France.

By going to France I would, of course, be forcing myself to struggle with whether or not to turn animals into dinner—one of the more controversial subjects we face as modern, first-world human beings. But it wasn't controversy I was looking for. I was looking for an unmediated, uncompromising honesty with myself, with my community, and with the world at large—an honesty I felt I'd lost over the course of the ten years I spent as a lifestyle magazine editor. In picking up a knife, I desperately wished to divorce myself from the world of magazine writing. I no longer wanted to write about the genuine article. I wanted to be the genuine article.

I also wanted to reconcile the opposing poles of my childhood—my dad's fishing boat, my grandpa's bright-orange hunting cap, my mom's sprout-and-cream-cheese sandwiches, my hometown's predilection for tempeh and tie-dye. As a child, I'd spent weekends hunting and fishing with my grandpa and dad, both genuine articles in their own right. I have memories of dragging dead pheasants through thick woods, dew-soaked deer carcasses swaying from trees, hooked fish knocked dead in the head with a small baseball bat we called the Fish Whacker—"to ease their suffering," Dad always said. But, much to my father's dismay, I'd turned vegetarian in my teenage years, mostly because everyone around me in the liberal college town of Eugene, Oregon, seemed to be doing it. By the time I met Tom and we moved to New York City, in 1999, I was eating meat again. By then it had been well over ten years since I'd climbed into my dad's fishing boat or eaten Great Aunt Helen's chicken-fried venison. I ate meat, but, like most Americans, I was completely removed from the processes that brought it to my table, even if, as a food writer, it was my job to ask where that steak came from. This remove nagged at me, but I did little to remedy it.

In so many ways, my entire life felt this way. The choices and decisions I made—to stay in a career that made me unhappy, to stay with a man whose vision of the future did not include me—felt rather more automatic than thoughtful, more convenient than meaningful.

SOMETIMES WE MUST CLIMB UP to a high perch and dangle our very identity—everything we've deemed certain and continuous and reliable and safe in our lives—over the edge. We must open our hands and let go of all of it, and then listen for the awful sound it makes when it hits the ground. Only then can we go in search of meaning.

I broke my own heart. I wrecked my career. I went to France to confront a reality I had, for most of my life, chosen not to. A black hole opened up in front of me and I stepped in. It was crowded and noisy and hard to see in that intimidating, liminal landscape, a landscape roiling with difficult sacrifices, certain deaths, untold complexities. But stepping into that landscape was one thing. Remaining long enough within it to rake its depths for meaning was another thing entirely. Black holes, after all, according to some theorists, have the potential to become entirely new universes. *But not without painful contractions and compactions followed by a blinding and expansive undoing,* the theorists warn, shaking their fingers at us, furrowing their brows.

"Why would a beautiful woman go and do something like that?" the man on the plane had asked me.

"It's complicated," I told him.

Somewhere over the Atlantic, a flight attendant asked, "Chicken or pasta?"

"Does the pasta have meat in it?" I asked.

"It's vegetarian," she said.

"I'll take the pasta, then."

The man with the wilted silk handkerchief raised his eyebrows. He had no more questions for me. I was beyond his comprehension. Like my name, it all took too much explanation.

THREE

While our pilot circled above Toulouse, waiting for a signal to land, I gazed out at the French sunset, which turned the city's buildings the color of fresh cream, the roofs all rust-colored ceramic tile. I'd read on the airplane that Toulouse is known as la Ville Rose, the Pink City, but the city was awash in a lipstick ocher, the sort of deep dusk color that appears on the horizon after a day that has seen both the sun and the rain. From above, the city was a humble, honest beauty, earthy and solid, to be sure, but sultry in her way.

Once I was down on the ground, walking through her innards, where pedestrians and bicyclists and men and women on mopeds ruled the streets, she expressed a much more frenetic, cosmopolitan bustle. Hailing a taxi, I noticed my hand was shaking, the physical manifestation of a nervous brain assaulted by an onslaught of new information.

I had one night in Toulouse before moving on to Agen, where I would meet my host, Kate Hill, an American who'd lived in France for more than twenty years and scratched out a living teaching people like me how to cook and eat and butcher the Gascon way. I'd met her only once, in Portland, but she hadn't even flinched when I called her up and asked, "Do you know anyone who could teach me how to butcher a pig?"

After checking into my hotel, I took advantage of what little time I had to wander the banks of the Garonne River alone.

But as I walked along one side of the river, I worried: surely, on the other side, something was happening that I should not miss. Hence when a blushed brick-and-stone bridge finally presented itself, I crossed over, pausing briefly to look down at the muddy brown water, which appeared, in that strange blue hour at the end of the day, as if it were not moving at all. Once on the other side of the river, I immediately felt as if I should have continued walking on the bank I'd been traversing before, and so, when another bridge presented itself, I crossed over again.

Perhaps this is how I should find my way to Agen, I thought: by following the Garonne, crossing every bridge I came to until I got there, flipping one side for the other. Maybe I'd keep following the river all the way to its salty mouth in Bordeaux. Now that would make a good story.

I no longer had to write about where I was, of course, but I couldn't shake the feeling that I hadn't done enough research before I arrived. In fact, I hadn't done any, and my fear was that I wouldn't have the *right* experience now that I was here.

Eventually I turned away from the river toward the city lights, realizing for the first time since I'd landed that I was alone in a foreign country, with no one else to answer to. I wasn't a journalist any longer. I didn't have to *look* for anything if I didn't want to. I didn't have to pass out business cards to every chef whose food I ate. I didn't have to order the most representative dish on the menu. I could even make terrible choices about what I ordered and there would be no consequences.

I turned a corner into an alleyway lit from above by bodega lights. The last time I'd wandered the streets of a city in France, I was with Tom in Paris. We'd sat down at a cheap bistro and ordered a carafe of tart red wine and an *onglet,* hanger steak. Tom, a classics major who could string together sentences in ancient Greek and who loved the tricks language can play on us, couldn't remember how to say "medium rare" in French. The waiter kept asking, *"Bleu? Bleu?"*—the word, we later learned, for "rare"—to which Tom kept saying, *"Rouge! Rouge!"* and laughing. The waiter didn't think it was funny. It'd been more than a year since Tom and I made the painful decision to leave each other. I'd supposedly moved on. But as I moved farther into the alley, I wished that Tom was walking next to me.

Save for an old man emptying a carafe of red wine at a bistro table by himself, I was alone in this alley. My footsteps turned suddenly loud and hard, as if someone had just cranked the volume up on whatever scene I'd walked into. The man sipped his wine and stared at me without blinking as I passed him. I made an awkward attempt to walk on the balls of my feet in order to quiet my step. Walking too hard had lost me my job. Who knew what other sorts of trouble it could get me into.

JUST A FEW MONTHS before I stepped into that alley in Toulouse, the new editor in chief at the city magazine in Portland where I'd worked for three years called me into her office.

"People in the office have been talking to me about you," she informed me.

"And what is it they're saying about me?" I meant to sound neutral, but my sarcasm immediately took over. This wasn't the first time I'd been called into her office to be told that so-and-so had a problem with me, although by this point I was pretty sure my editor in chief was making most of these problems up. "Pretty sure" being the key phrase.

Sometimes I left her office convinced that I *was* in fact a conniving contrarian on a mission to overthrow her magazine government and undercut anyone who got in my way.

"They're saying that you walk too hard," she said, in that apologetic tone people use when they tell you a piece of lettuce is stuck between your teeth. "And, frankly, I have to agree with them." Her little black Scottie dog, Mimi, who was recently rumored to have urinated on our staff common-area rug, yipped as if to emphasize her point.

"You've got to be fucking kidding me." Ever since the new editor in chief's recent arrival, most of my colleagues at the magazine had barely been capable of questioning a headline in front of her, let alone lodging a complaint about the tenor of my step. But maybe they *had* offered this information freely. Maybe I *did* walk too hard.

"I'm not kidding," she said. "I need a managing editor who doesn't stomp around all the time."

I stared at her in silence for longer than one is supposed to in these situations, waited until her face turned a bright, splotchy red, then left without saying another word.

"She hates me," I said to my friend and colleague Jill, the magazine's second-in-command, with whom I'd just spent the past year voluntarily working long hours to hold the magazine together without an editor in chief and with no pay raise.

"I'm pretty sure she hates me, too," Jill said.

"I think she's going to get rid of me," I said. "And you know what? I don't even care if she does."

It was true: I didn't have any energy left to care anymore. I was done. It'd been a shit year. After I'd moved out of the house Tom and I shared, my therapist diagnosed me with depression and put me on Celexa. On top of that, I was drinking a bottle of wine a night and chasing it with

Ativan just to get to sleep, but I rarely stayed asleep for long, because I'd repeatedly begun to wake up in my lonely World War II–era apartment with the arched doorways and wooden floors that echoed when I walked on them, convinced that someone was standing over my bed, looming over my sleeping body. My therapist tried to convince me it was my childhood self attempting to contact me. I, on the other hand, knew it was a malevolent ghost who wanted to bash my head into the wall.

"Why does the ghost want to bash your head into the wall?" my therapist asked.

"Because suffocating me wouldn't be shocking enough," I remember saying.

When I wasn't dreaming of menacing night visitors, I dreamed of teeth. Broken teeth. Rotting teeth. My teeth transformed into fangs that jumped out of my mouth and bit my face.

I read recently that the part of our brain that registers the excruciating pain of a very bad toothache also lights up when we are heartbroken. Losing Tom felt like having someone yank out all of my teeth with pliers when I wasn't looking. The loss left me howling, ferocious for months—nay, years—after. I believed my night visitor was on the hunt for more teeth, even though my mouth was empty now, save for my tongue, which obsessively ran itself along my mouth's interior in an attempt to master this new, devastating absence.

After Tom and I left each other, I'd stayed single for a short while, but I eventually began sleeping with Will, a kind, enigmatic, handsome furniture maker, with an impressive grasp of poetry and metaphor, who also had mold growing on his bathroom ceiling and a chewing tobacco habit. Falling in love with Will was kind of like getting dentures—which is a terrible thing to say, but I mean it in the most complimentary of ways. I was relieved to have a sturdy, reliable set of teeth back in my mouth—I felt the fit was quite snug—and the phantom pain mostly gone, but at the end of each day, I stood in front of the bathroom mirror and remembered that these teeth were merely a simulation of the ones I'd had before.

Will was kind and loving in his way, but in time I knew his love for me greatly outweighed my ability to love anyone, including myself. What I secretly wanted—without really ever being able to admit it to myself—was to quit everything and everyone and descend into a deep, dark hole. Being unemployed, accountable to no one, doing nothing, sounded like bliss.

TWO DAYS LATER, our editor in chief called Jill and me into her office and said she was letting both of us go, faking tears the whole way, telling us what a hard decision it had been, blaming it all on the recession.

Laid off. Fired. Whatever you want to call it, my magazine career was more or less over, unless I moved back to New York, which I was in no position to do. The dour, pale woman who served as our magazine's lone human resources administrator instructed us to grab our things, monitoring us as we did so. I scanned my office and instinctively grabbed the apple sitting on my desk. *Might as well start stocking the pantry now*, I thought. *Although I might not even have a pantry by the end of the month*. I'd done the math the night before.

"It's time to go," she said, escorting us as a security guard might through the staff cubicles and down the stairs to the front door, just in case one of us tried to, what, punch someone? We were outcasts now, voted off the island. Dangerous elements. I stared straight ahead, looking at no one, and walked as hard as I possibly could.

As my feet hit pavement, I could feel each hard, calcified, bitter bone in my body begin to soften. I was shaking and full of angry adrenaline, but I also felt relieved.

"Thank fucking God," I said to Jill. "Let's get drunk."

And we did. Or I did, since Jill was six months pregnant. At Dan and Louis Oyster Bar, I proceeded to down five old-fashioneds in less than an hour. It was only ten in the morning, but our colleagues each found an excuse to step out of the office and join us. I made a lot of jokes. I laughed. I bumped fists, high-fived everyone, and triumphantly toasted new beginnings. But hiding within was a creeping sense of total and utter loss.

As I HEADED BACK toward the center of Toulouse, a deep fatigue set into my legs and rose up into my chest. My stomach felt as though it were eating itself from the inside out—nerves, probably, and hunger, certainly—but since I was not required to eat for a story assignment, I defiantly skipped dinner. This was a small act of defiance, but because it went against what I believed constituted my core nature, it inspired in me a familiar existential panic.

Who was I? Where was I? How had I lost myself so completely in a relationship that clearly wasn't going to work? And how had I lost myself in a job that had me sacrificing sleep for . . . pulp? Magazine pulp. "Ten Top Restaurants." "Twenty Cheap Eats." "Five Ways to Eat a Mango." Pulp that people hardly read while eating bacon cheeseburgers at McDonald's or sitting on the toilet. Pulp made out of the same pulp we made the year before, and the year before that. How had I remained so enamored with this pulp for so long?

That night in Toulouse, I slept for fourteen hours in a hotel room that smelled like a grandmother's perfume. Its tasseled curtains and rosy wallpaper conjured a retired flapper's boudoir. The June air pulsed with static heat. The sole window in the room refused to open, so I slept fitfully, rolling from my left side to my stomach, then, ten minutes later, to my right side, then to my back, then ten more minutes and I was on my left side again, as if my body were affixed to a rotating spit.

FOUR

For breakfast, at the train station, I ordered, via pointing and a few *merci*s, an apple croustade, a coffee, and an Armagnac. Early on in my days as an editor at the national food-and-drink magazine *Saveur*, I'd fact-checked a few sentences about Armagnac, so I knew that it was a brandy made from white wine grapes that's unique to Gascony and, unlike Cognac, is distilled once instead of twice, therefore requiring more time aging in oak barrels. But I'd never actually tasted it. I guessed that Armagnac's flavors would meld well with the croustade's vanilla-laced, buttery dough, the tartness of the apples, and the bitterness of the coffee, and I was right. The amber-hued Armagnac tasted of butterscotch and vanilla at first, but it also possessed a musky acidity that got along well with the rest of my Gascon breakfast. The creamy burn of the drink on my tongue thrilled me. Instead of sitting at a desk fact-checking Armagnac, here I was in Gascony, drinking it. This was progress.

At the gift shop, I bought a postcard for Will. On the front of the postcard, an older gentleman in a beret, who looked a good bit like the original Ichabod Crane, with wild, crooked eyebrows, had managed to stuff his bulbous red nose into the tiniest of Armagnac snifters. On the back of the postcard I wrote: "My first breakfast in Gascony. Wish you were here. I'll be back soon. Sorry I moved in and then ran away to France, but I'll be so happy to see you when I return." I meant it, but I also never sent the postcard.

I took a train one hour northwest to Agen, past signs for towns with names like Moissac and Montauban. I loved the dominance of vowels here. Open-mouthed sounds. Out my window, the Garonne Valley passed by. With its rolling pastures and grasslands interrupted by small parcels of forest and orchards, it looked so much like the central Willamette Valley, where I spent my childhood in Oregon, in the little town of Alvadore. I could see three main crops thriving here, which I would later learn were sunflowers, corn, and rapeseed, alongside wheat, barley, and rye. Sprawling white roses, chicory flowers, Queen Anne's lace, and elderflowers grew wild along the side of the road.

My French was terrible—the four weeks I'd spent the month before in a 101 class at Portland Community College had not really stuck—but I entertained myself anyway by making up meanings for the business signs we passed. I saw one sign above a business with a word like *salvage* or *salage*, next to the word *cuisine*, but it wasn't the word *sauvage*, which I knew meant "wild." I decided that this particular business worked to salvage wild food. Could something wild be salvaged?

I GREW UP hunting and fishing in the *relative* wild with my dad and my grandpa. We fished for bullheads and bluegills in the Cottage Grove Reservoir, using worms and Velveeta cheese that we rolled into tiny balls and sprayed with WD-40. Sometimes we'd drive three hours south into the wilder landscape of the Umpqua River to hook steelhead with woolly bugger flies. We wandered dense forests with Gabe, our yellow Lab, who'd run ahead of us to scare pheasants out of the underbrush. We set up duck and goose decoys in the dry fields and shallow lakes near Crane Prairie. Whenever Dad blew into his duck whistle, he sounded like a cartoon, and my twin brother, Zach, and I had to cover our mouths to stifle our laughter.

"Stay quiet," our grandpa, Dutch, would gently scold us in a whisper. "The birds are coming. Can you hear them?"

I suppose I lived a wilder childhood than most people I know. The modest house in Alvadore that my dad, a carpenter, built for my mom, my twin brother, and me sat at the end of a short, potholed dirt road—the signage read PRIVATE ROAD—which branched off of another potholed dirt road called Fruitway. We were surrounded by fields of tall grass and weeds, tangles of blackberry vines, and an abandoned plum orchard, where the old plum-picker shacks were slowly collapsing in on themselves. We roamed Fruitway chasing our neighbor's ducks and geese, feeding lettuce to Marie's pygmy goat Charlie, stealing eggs from underneath the half-dozen laying hens at Tony and Buck's, holding long conversations with Rosy and Wayne's sheep. The farm animals in our particular neighborhood—and there were a lot—were not pets, exactly, but they didn't appear to have any particular utility, either. Rather, they seemed to serve as reminders of a long-ago time when the people of Fruitway Road worked the land and raised animals for food for themselves and their neighbors—back before everyone found jobs in the city and began buying their fruit and vegetables and meat at a big grocery chain in Eugene, thirty minutes away. By the end of summer, mean old Ratliff's U-pick strawberry field, which sat at the end of Fruitway Road, smelled of sweet rot.

The Willamette Valley was and still is considered a thriving agricultural community, with vineyards and grass-seed farms, hazelnut and fruit orchards, and plenty of successful vegetable and berry operations.

But on Fruitway Road, we lived a simulation of country. We acted country, but it wasn't *really* country. Not anymore. We didn't need to look very far to find evidence—rusted tractors, expansive trailer parks, hay bales left to spoil in the field—of the gradual disappearance of the agrarian way of life.

The green-and-tan fields of the Garonne Valley that I saw out my train window, however, had recently been tilled and planted, the limbs of apple trees expertly pruned and netted to protect them from hail. We passed multiple farm equipment stores, the parking lots full of men in muddied blue coveralls and worn work boots. Gascony looked alive with new growth and old rhythms, both wild and controlled at once, its land still deemed useful, bounteous.

IN MANY WAYS, a longing for the "wildness" and "country" of my childhood had inspired me to travel to Gascony in the first place. But not before I spent the first months after losing my job attempting to emulate, as best I could, those circumstances I'd thought had been so continuous and predictable in my adult life: a man in my bed, a certain kind of work.

No longer able to afford my apartment, I'd moved in with Will, who had generously allowed me to turn the dingy back room of his house—which, over the years, he'd filled with old filing cabinets, antique guns and bullets, gargantuan concert-worthy speakers, and empty chewing tobacco canisters—into my office. I planned to make a go of it as a free-lance magazine writer. What other way forward was there? I had Will. I had my ten-year writing-and-editing career behind me. If I just held on to what had once been reliable in my life, surely it would all work out.

But after hustling a few story assignments from editors in New York and Portland—an as-told-to for *GQ* about the perils of raising chickens in your backyard, a roundup of the best tamales in Portland for *The Oregonian*, our state newspaper—I stopped completely. I stopped calling editors, stopped pitching stories, stopped turning my computer on each morning and pretending I was excited. When the paychecks arrived, I waited days before cashing them. I questioned, suddenly, whether writing these stories was an honest way to make a living, and so I felt guilty accepting the money, even though I desperately needed it.

Instead, I sat for days staring out my new office window onto Will's backyard, which we'd transformed, with a roll of sod, from his dog's muddy, shit-strewn, ten-by-ten strip of toilet to a bright-green lawn. I busied myself watching opportunistic robins hunt for their next meal. How had the worms arrived so quickly? Had they come with the sod? Had we paid extra for them?

The robins stood still, cocked their heads, waited. If a worm did not make itself known, the birds scurried a few inches away, cocked their heads again, waited. What did they hear? What invisible movement in the earth were they able to detect that I could not? In their dedicated search for their underground prey, the robins must have had to quiet their hearts to a barely discernible pulse, to hush the whirring of their tiny bird brains to not even a whisper. When a robin finally retrieved a writhing, juicy worm, I'd stand up out of my chair and press my face to the window to watch the robin eat, in just a few swallows, its entire length. How quick and decisive her service to this desire appeared to me. It had been so long since I listened to my own desires that I would have to observe many more robins pulling many more earthworms out of the ground in my new backyard before I could even discern their outlines.

There was no avoiding it. Like a pendulum, the impact of the preceding year had swung away from me just long enough that I'd been able to settle a few important practicalities, like where to live and how to feed myself, but the pendulum was on its way back toward me and moving fast. Loss is a round-trip traveler. When we push her away, we only lend more momentum for her return trip.

PERHAPS WRITING WASN'T my thing, I told Will; maybe I needed to do something different. Whatever it was going to be, I needed to reinvent myself, and fast. But as I waded through the classifieds, I found that the kind of job I craved, the kind of reinvention I began to envision for myself, would take me several steps *down* the career ladder I'd been climbing for so long.

The work I began looking for, without even understanding why— the work that I felt, curiously, better cut out for—was the kind of work I'd originally come from. Cue picture of my brother and me posing with our first rainbow trout, cue picture of us under the hood of my dad's truck, helping him fix the engine with our plastic toy hammer and wrench. Cue Dad and Grandpa and Zach and me in camouflage jackets, with bows and arrows and fishing poles. Cue Dad and me bending rebar for the foundation of the third house he built for us to live in. Cue Grandpa in his garden. Cue Grandpa and Grandma canning pickles and tomato sauce. Cue my very first pocketknife. Cue my mom and her mom sewing their own curtains and clothes, baking pie from scratch. We built our own houses. We fixed our own cars. We grew our own vegetables. We made our own pie. We even harvested our own meat— but only sometimes, given my mom's disgust at the sight of a freshly killed Canada goose. I came from a family of workers, mostly of the blue-collar variety, with a little administrative flair thrown in, who knew how to survive by working with their hands. I wanted to *be* the robin pulling worms out of the earth, not the person standing behind a window *watching* the robin pulling worms out of the earth.

My favorite book as a kid had been Scott O'Dell's *Island of the Blue Dolphins*, about a young American Indian girl who is mysteriously abandoned by all of her people on an island off the coast of California. Drenched in loss though she is, she continues to live on the island, alone, eventually befriending a wild dog, her constant and only companion. In time, she becomes her own kind of wild. She assumes traditionally male tasks out of necessity: hunting and fishing, making spears, carving canoes. She is alone, on an island, living in a house made of whale bones, surviving on abalone and devilfish and berries that she harvests herself, with an animal as her only friend. I remember thinking that this sounded like a pretty great life, even if, in the end, a bunch of white people showed up and forced her to shed her cormorant feather dress.

I read that book right around the time Aretha Franklin and Annie Lennox sang "Sisters Are Doin' It for Themselves," my favorite song back then. I was raised on *Ms.* magazine, with a fishing pole in my hand. When I was seven, my mom took me to a Democratic Party brunch to meet Geraldine Ferraro, who was running for vice president. My dad taught me how to hold a hammer when I was five. At the age of nine, he handed me a dead trout and my first pocketknife and said, *Here's its spleen, here's its heart, here's its gills, here's its intestines.* As a young girl, my particular feminist ambition fantasy was to live alone on an island and survive by using my own two hands and doing things girls weren't often taught to do for themselves.

In time, I turned away from this fantasy. I'm not sure whether these things are related, but when our parents moved us from the rural town of Alvadore to the city of Eugene—really more of a small-town haven for like-minded hippies than a true, diverse city—when my twin brother and I morphed into moody teenagers in the late 1980s, when my dad and my grandpa stopped hunting and fishing together, when I became a vegetarian, that's just about the time I remember shifting my attention from the wild, countrified, physical, natural world of my childhood to a more urbanized, abstracted world inside of my head. I gave up the violin for theater. I discovered the Steppenwolfs and Siddharthas of Hermann Hesse, the roadside cowgirl philosophers of Tom Robbins, the defiant Sulas of Toni Morrison, and I started thinking about what it meant to be a writer. This was also when I turned my back on the sheep and geese, the turtles and goats and bird dogs of my childhood, the fishing poles and bows and arrows, the fish heads and spleens, the blackberry bushes and cherry trees and abandoned plum orchards, and all the difficult real-world truths that they presented. We were city dwellers now. And so, in my move from country to city, I turned away from the real world of real things, perhaps in an attempt to find meaning in my distance from it.

Yet, in the back of my mind, I continued to believe that I had immediate and inherent access to real-life skills, like starting fires and chopping wood, like gutting fish and killing animals for dinner—"survival skills" is the phrase we use nowadays, since, at least in the industrialized world, most of these needs are taken care of for us no matter our class status, and, presumably, only in a total emergency would we need to rely on such skills ourselves.

In reality, I'd spent little time mastering any of these skills. I hadn't needed to. If I were really lost in the wild, I'd actually have little idea how to feed myself. I'd be alone and cold and hungry and then probably dead in that *Island of the Blue Dolphins* fantasy of mine.

Maybe, somehow, after losing my job, I was feeling a little lost in the wild, or, rather, lost in my particularly urban notions of the wild. Sure, I could spin a good story about being lost in the wild, but I didn't really know how to *be* lost in the wild.

Maybe these thoughts were unnecessarily nostalgic. But what is nostalgia if not a glance toward the past in order to alight upon a workable vision of the future?

FIVE

My train arrived in Agen a bit early, or maybe Kate Hill was late, and I waited with my luggage in the parking lot of the small train station with a mix of adrenaline and self-doubt. Was I really doing this? After a few minutes, Kate drove up in a beat-up silver Peugeot, got out of the car, and strode toward me in a pair of bright-green Crocs, a billowy white linen sundress, and her signature indigo-blue do-rag.

"*Bonjour!* Welcome!" she said, positively beaming, giving me a kiss on each of my cheeks, followed by a warm, lingering hug. "Are you hungry? There's a great little café in the train station. Let's go there before we get in the car."

We sat down at a small bistro table by the bar. Jet lag still had a grip on my brain, so I let Kate do most of the talking.

I'd reached out to Kate about a month before I arrived in Agen, on the recommendation of Robert Reynolds, a much-loved Francophile chef and underground culinary mentor and teacher to many of Portland's cooks, who often took his students to France. After I wrote a story about Robert for the city magazine, we'd kept in touch and eventually become friends, after which he'd regularly invite me to his famous foie-gras-and-bubbly parties to honor some culinary whiz or another who was in town from a faraway land to share stories and cook with Robert. One winter evening, Robert's guest of honor had been Kate, a cookbook author and teacher who'd swung into Portland all the way from her home in Gascony to teach a cassoulet class with Robert. As we talked over glasses of rosé, I took to Kate almost immediately. She had a way of becoming familiar quickly, as in she already felt like family. Hers was a big presence—not in an egotistical way, just in a big-life, big-thinking kind of way. The constant verbal and collaborative unfolding of ideas served as her fuel, and so she came off as a supremely social being, dependent on the presence of others to complete her thought processes and put those thoughts into action.

That night at Robert's, she spoke to me of a family of pig farmers and butchers in Gascony who "did pork the right way," of a beekeeper who was teaching her how to produce honey, of a goat cheese maker and a foie gras duck-and-goose farmer whom she often took her paying students and visitors to meet.

I told her I was working on a story about *bavette*, a French cut of beef I'd just tried at a new local restaurant I was reviewing, and that it had been difficult for me to get anyone at the restaurant, or any butcher shops in town, for that matter, to tell me what part of the animal the cut had come from, or why *bavette* tasted so good. The steaks had come to the kitchen already cut, vacuum-sealed in plastic, sent from some processing facility two states over and delivered by a distributor, so the chef hadn't really had to think much further than that.

"Whole-animal butchery is a lost art in America," she said, shaking her head. "It's all done in big processing facilities behind closed doors."

I'd gotten some sense of this several years earlier, while helping *Saveur*'s food editor, Melissa Hamilton, work on a story about an old-school Italian American butcher shop in New Jersey that she'd grown up going to. The owners of the butcher shop, the Maresca brothers,

Emil and Joe, sourced whole animals from local farms and did all the killing and cutting of those carcasses themselves, something very few "butcher" shops did anymore, opting instead to buy precut meat from distributors who bought from factory farms. When I spoke with Emil and Joe by phone, I sensed I was speaking to men who were mournful that their lifelong trade and knowledge would not be passed on. They were a dying breed who spoke defiantly of grocery chains, big-box stores, and confined animal-feeding operations as if they were meteorites falling from the sky, headed straight toward their well-oiled wooden butcher block. I knew little to nothing about butchery, but Melissa and I had talked, very briefly, about the idea of taking over the shop together when the brothers retired. "No one else is going to take it over," I remember Melissa saying wistfully.

"On the other hand," I said to Kate, "a few chefs in town are breaking down whole sides of pig on their kitchen counters." I'd recently noticed that a growing number of chefs in Portland were teaching themselves butchery and putting things on the menu like pig-head pâté and sweetbread salad. Shortly before I was laid off, I'd helped produce a photo shoot for our annual "Best Restaurants" package, featuring an up-and-coming chef named Gabriel Rucker, who'd just opened Le Pigeon. For the shoot, he'd agreed to pose for us sitting next to a pig head at one of his dining room tables. Later, as the photographer and I sat sifting through the resulting shots, I'd stared at the pig head in those photos and thought, *There's something important there,* without really knowing why I felt that way. We lost at least a few subscribers because of those photos.

"It's funny how pigs' heads and sweetbreads are considered novelty here," Kate said. "In France, they're a way of life."

At the time, I didn't really understand what she meant by that, but I wanted to find out. What would it mean to have pig heads and sweet-breads be a way of life, not just novelty? How would I eat differently? How would I think about meat differently?

And then I surprised myself by blurting, "I kind of secretly want to be a butcher."

"You should come visit me in Gascony, then," Kate said, and I thought, *Yeah, maybe someday. Maybe in another lifetime.*

After that, whenever I saw Robert, he'd nudge me. "So when are you going to go to France to become a butcher?" he'd ask.

"Just as soon as I quit my job," I'd say, joking. "Any day now."

Two years later, while watching robins hunt for worms in a back-yard that was not my backyard, I called Robert.

"Hello, dear," he said.

"I'm ready to go to France," I told him.

"I'll connect you with Kate," Robert said. "You remember her, right? You met her in my kitchen. She'll help you. Now, when are you coming to dinner?"

You met her in my kitchen. All the people I want to meet always end up in the kitchen.

"Of course I remember her," I said. "She's hard to forget."

I wrote to Kate. *I have no money,* I said. *But I want to come to Gascony and learn butchery. I'm a carpenter's daughter. I'm willing to work. I'll paint your fences.*

Kate didn't even flinch. *Of course you want to come to France to learn butchery. Of course I will help you. Of course you can do manual labor for me. Of course I will put you to work. I will feed you, too. House you. Take you to markets. Teach you to make charcuterie. We can visit Armagnac makers, too, if you like. And goat cheese makers. And bread bakers. There's a family of pig farmers and butchers, the Chapolards. They own a seed-to-sausage operation. They'll teach you when I can't. You'll work in their cutting room. You'll go to the abattoir and work the market with them. And there's a duck-and-goose farmer named Jehanne. She wins gold medals for her rillettes and foie gras. You should meet her. We need to talk about resources, financial and otherwise. Ways to generate income to pay for pigs and pedicures. It's doable. Let's put our thinking caps on.*

Pigs and pedicures. Was Kate even real? Had I imagined her?

She was very real. And she was serious about taking me in, in trade for work and a very small amount of money, Tom Sawyer style. A week later, I bought a plane ticket.

"I'm sorry," I told Will. And I was sorry. Sorry to not be able to contribute much cash to our new household, but to have found an un-used and not-yet-expired credit card in the back of my filing cabinet to pay for this trip. Sorry to have moved in, only to disappear a few short months later.

"I want you to be happy," he said. "Just don't come back from France and leave me."

"I'd never do that," I said. "I'd be a terrible person if I did that," and I meant it.

He bought me a ring that said YOURS on the inside. And as a parting gift, he gave me a palm-size silver figurine of a fat pig. On the side of the pig he'd carved the crude lines of a butcher's chart. On the belly he etched the words PORK DRUNK LOVE. It felt like a proposal of sorts, one I accepted without ever actually saying the word *yes*.

And then I was gone.

AT THE TRAIN STATION CAFÉ, Kate bought us two glasses of red wine and a plate of steak frites, *à point*—medium rare. In the States we're often charged for this particular meal at restaurants as though it were fancy. Here it was fairly commonplace, like a burger or a ham sandwich, inexpensive and also delicious.

"This week," she said, "I'll take you to a few different markets so you can get the lay of the land. Maybe tonight we'll make something with foie gras. And it's already cherry season, so maybe we'll make a cherry clafoutis."

"Maybe tonight we'll make something with foie gras" probably sounds fancy. I think maybe it did to me, at the time—I had always felt fancy going to Robert's foie gras parties. What I didn't realize was that while Gascons considered food like foie gras (fattened goose or duck liver) to be special enough that it should be eaten with reverence, they also considered it simple, good food. And simple, good food was in abundance here because people worked hard to produce it. Although various foie gras dishes, as well as many other traditional Gascon reci-pes, have been co-opted by sophisticated urban restaurants that charge hefty prices for them, these were born of the Gascon countryside. But I didn't know this yet. And so, when I would describe these dishes to others back home, or send photos of them, I got a lot of "ohhhh, how fancy" comments from friends and family, and I did not question their judgment.

I devoured my steak—which I told Kate tasted a good bit more like actual meat than any steak I'd had in the States—and washed it down with the red wine, which captured the flavor of that dirty red ocher of Toulouse in all the right ways. Then we crammed my overstuffed suit-case into the tiny trunk of Kate's Peugeot and crossed over the Garonne River once again, past the tiny town of Brax, over the invisible bound-ary that makes up the panhandle borders of the village called Sainte-Colombe-en-Bruilhois (pronounced *Sawnta-Cohlome-on-Broolwah*—oh, the glorious vowels!), down a narrow, potholed road lined with fields of sunflower stalks and apple orchards covered in gauzy netting, into a tiny hamlet—a speck of a dozen houses on a map, a Gascon version of the Fruitway Road I grew up on—known around these parts as Camont, to the sturdy stone *pigeonnier*, a beautifully haphazard structure whose ragged walls contained layers of broken brick, coarse mortar, smooth river rock, and jagged scraps of limestone, which Kate had converted into a rustic cooking school and guesthouse, set on the Canal du Midi. After years of living on a river barge and captaining it through the canals of Europe, this is the place where Kate had anchored her barge, the *Julia Hoyt*, for good in 1989. This same barge had brought her to Gascony, helping her find the friends and mentors who taught her how to cook the Gascon way, lessons she would now, I hoped, impart to me.

As we pulled into Kate's gravel driveway, her gargantuan mutt of a dog, Bacon, greeted us by barking and jumping up and down. His bloodshot brown eyes expressed guilt and annoyance simultaneously. His big mouth, with a dog beard hanging from it, gaped open, panting. Large patches of black wiry hair interrupted the rest of his dusty white coat.

"Bacon. Sit. *Couché. Couché*. Sit. Lie down." Bacon kept barking at us. "He's such a big baby," Kate said, turning to Bacon and scratching him behind the ears. Kate explained that he had come from the Chapolards' pig farm. "The Chapolards say he's half pig, half dog." That seemed about right. He reminded me of photos I'd seen of old pig breeds.

Everywhere I turned, I encountered a different garden, each blooming with distinctive personalities. Bright-pink and purple hydrangeas. Climbing nectarine-orange roses. Bay laurel. Spirea. Sage. Rosemary. Thyme. Human hands had planted everything in front of me, but the entire place felt savage in its beauty. The mustard greens had already gone to seed. There were carrots to be picked, lettuce and nasturtiums that still needed to be planted.

"So," she said to me after I entered, for the first time, her earthy yellow kitchen, with its red stone fireplace and matching stone floors, cracked and worn by the feet of her many guests, after she'd put on water to boil and poured us a special tisane made from dried mint and lavender from her garden, "you've got seven weeks in Gascony. What do you want to learn?"

And then, there I was cooking foie gras in the kitchen with Kate, buying duck prosciutto at Gascony's markets, stomping around the Chapolards' pig farm, negotiating whole sides of pig in their cutting room, wandering the stark corridors of the abattoir, face-to-face with the real, honest-to-God, genuine article.

PART 2

SIX

Before I stepped onto the Chapolards' farm, Dominique, one of the four Chapolard brothers, and his wife, Christiane, wanted to have me over for a proper dinner. Plus, Kate wanted to make sure I was well steeped in Gascony's market stalls, butcher shops, and *boulangeries*. And so a week went by before I even picked up a knife or met a pig.

Getting oriented inside of Kate's world meant, first and foremost, entering into her particular brand of well-cared-for, formalized wildness, acclimating to her fast-moving brain, her unpredictable rhythm, her complex social network. Kate was a passionate improviser whose boundaries between home and work were nearly nonexistent. Her kitchen served as her central office, but her work territory stretched from her kitchen to her outdoor dining table, then out farther to her *potager,* where we harvested fresh herbs, beans, tomatoes, radishes, carrots, and bitter greens, to the chicken shed, where we gathered eggs and occasionally harvested roosters and ducks, then to her neighbors' down the road, the Sabadinis, who, nearly every winter since she had moved to Gascony, had invited her to help them kill a pig and turn it into pâté and sausage. And out, farther still, into the market stalls and farms and dining room tables of the neighboring villages, where she met people like the Chapolards, who first taught her that a good *boudin noir* should always be cooked and served with apples. All of these people were farmers and growers, cooks and gardeners, passionate eaters, too, carrying on the traditions that had been passed down to them. They were Kate's extended French family, a family perpetually sitting down to an everlasting meal together. "It's a life, not a lifestyle," Kate often liked to say.

So how did Kate make a living, a life? With either money or time and skills, people from all over the world paid Kate for a seat at her endless feast. We paid Kate to introduce us to, as she liked to say, the butcher, the baker, the Armagnac maker. The goat cheese maker, too. The beekeeper, the winemaker, the bean grower, the strawberry farmer, the foie gras producer. Which meant that Kate's house was filled with lost people like me, who had little money, a lot of time, and a willingness to work for food, as well as people not like me—people with plenty of money but little time. For some, a stay at Kate's place was simply a break from their lives. There were plenty of high-maintenance empty nesters who simply wished to drink good wine and tell their friends about their first taste of *true* foie gras. For others, visiting Kate signaled the beginning of a life change. There were young women looking to open their own bakeries back home. Lawyers who had quit their jobs to start farms. Burned-out sous-chefs. Culinary school dropouts hungry to learn about food beyond hotel cuisine. Couples on honeymoons looking to bond over cassoulet. People searching for a way of life that felt, for whatever reason, impossible to find back home. Refugees from a land where the meat was bland and the bread had no nutrients. All of us in search of flavor, knowledge, connection.

Kate helped us find all of this. She was our foie gras whisperer, our *jambon* translator, our cassoulet queen. She schooled us in the Gascon way of happy hour by handing us a glass of sweet Floc de Gascogne, a *vin de liqueur* made from wine fortified with Armagnac, and a plate piled high with thick slices of *saucisson*. She taught us how to cook by taking us to the market and asking, "So, what's for dinner?" The answer should always be: whatever is in season, grown by whomever you can trust.

DURING MY TIME at Kate's, I wasn't the only one working in exchange for food, knowledge, and a place to lay my head. A man my age, a sous-chef named Jonathan, had come from Portland via Robert Reynolds, too. At first, Jonathan was standoffish. When he wasn't raking leaves or mowing the lawn or weeding or cleaning Kate's barge, he lay in one of her hammocks, reciting French to himself. Once or twice a day, he'd saunter into the kitchen and begin silently chopping whatever needed to be chopped. Together we planted seeds, washed dishes, spoke cordially to each other. He, like me, wanted to learn the art of Gascon butchery.

Jonathan and I, along with a few other lost souls who came and went during my time there, were Kate's summer entourage, moving from bedroom to barge to hammock at night, depending on how many clients Kate was hosting. We knew how to fit into the gaps Kate opened for us. Every once in a while, we probably drank more of her Armagnac than we should have, but she was kind in her scolding.

Kate's house was as eclectic as the clientele and friends and boat girls and garden boys she attracted. By "house" I really mean a sprawling indoor/outdoor complex of outbuildings, sheds, barns, trailers, her barge, gardens, patios, pantries, and lawns that I came to call Kate's French Commune. Inside her renovated stone *pigeonnier*—built in the 1790s for roosting pigeons and doves, once commonly kept for their meat and eggs and the fertilizing powers of their guano—where Kate taught and hosted students and guests, there were crocks full of home-made vinegar. A chandelier with dried hams hanging from it. Stacks of cookbooks along the stairs. A baby-blue bedroom—my room whenever guests weren't sleeping in it—with a claw-foot tub, a bed, a writing desk. Above the blue room, a sunny yellow bedroom with two beds and a Rapunzel-let-down-your-hair window. The walls and floors were made from thick stone, but the sounds of clanking pots, sizzling leeks, boiling soups, and laughing guests snuck in to greet me through the crack under my door. I could hear everything, all the time.

I came to think of Kate's kitchen on the ground floor of the *pigeonnier* as a character itself, singing a song of abundance: stacks of pots and pans, tins of tea piled ten high, wineglasses, Floc glasses, water glasses, Armagnac glasses, ramekins, cutting boards, tiny clay vessels meant for serving olives. Hand-cranked pepper grinders. Wooden bowls, each filled with a different kind of salt. Dried herbs hanging from nails. Handmade *cassoles* overflowing with bright-red tomatoes and runner beans and buttery lettuces. Drawers in the tiny fridge filled to the brim with stinky charcuterie and ripe, oozing cheese.

Outside, baskets hung from worn limestone walls. Six different kinds of roses ran along the pathways. An old circular wooden table that seated eight, perched underneath a sprawling arbor tangled with grapevines. A pull-along trailer Kate permanently parked and transformed into her office for the summer. A duck corral. A toolshed. A cold storage room she called her piggery, where she dried hams and duck prosciutto and stored her sealed jars of duck confit and jams and beans. Ceramic sculptures. Laundry hanging from lines strung between the *pigeonnier* and the barn. Hammocks strung from trees. A boules court. A black cat drinking milk from the barn window. An orange-and-white cat watching all the bustle from beneath a spirea bush. Kate's well-loved, fully furnished barge, anchored out back in the Canal du Midi, where Kate slept, and sometimes Jonathan and I did, too. An outdoor wood-fired oven perched atop cement blocks set atop an old metal table. Rusted wheelbarrows in which we'd build fires. A barn full of suitcases and more books, ratty umbrellas, old letters.

I recently befriended a woman who, when I am feeling pessimistic and down, likes to say to me, "I believe things will work out and they always do. I live my life in abundance. You should try it sometime."

Kate most definitely lived her life in abundance. She was a collector of people, objects, ideas, and ingredients. Everything, everyone, belonged until proved otherwise. There was always someone to talk to, always work to be done. There was always aperitif hour, too. Always a long lunch, a leisurely dinner. Always time for bike rides along the canal, and for swinging in the hammock.

"We're on Gascon time," Kate often said, and by that she also meant we were on Kate time. After ten years working long, disciplined magazine days without much rest, it took some getting used to.

By my third day at Kate's, I was stripping all the sheets from the beds in preparation for her next guests. Sweeping the floors, cleaning toilets. But then there was a need to head to the market and stock up on food for the days of entertaining and teaching that lay ahead. So we were off, driving the back roads—it never felt like there were front roads in Gascony—to Agen, to fetch whatever the market gave us. My first week it was fava beans, cherries, white strawberries, white asparagus, peas, a melon or two. Then we moved on to green asparagus, red strawberries, some early tomatoes, and all manner of soft, buttery, purple and green lettuces, some bitter and some sweet.

We briefly stopped at an outdoor market under a raised concrete parking structure in Agen, where Kate located a man with cages of live chickens and roosters in the back of his parked truck and began talking to him in rapid French.

"Next week I'll buy a couple chickens from him," she told me. "Our quarterly fox attack just cost me a few."

After the outdoor market, we drove to an indoor market, where a man with a timid smile, standing in front of a mound of melons, nodded at us. I picked a melon up to smell it, but the man said something curt in French and took it from me. He set it back down, wagged his finger, and picked up another one, looking to Kate, because clearly I did not know what I was doing.

"It's his job to tell you which melon is the best for you," Kate said. "It's an insult to pick up the melon yourself and inspect it, because it means you don't trust him."

Kate turned to him and spoke a few polite words in French. Without letting us touch or smell the melon he'd chosen, he gently placed it in a bag, as a mother might place a baby in a bassinet, and handed it to me, bowing his head slightly. Kate gave him a few coins.

"He asked me when we thought we would be eating it," Kate said. "I told him tomorrow, so he picked one that would taste good tomorrow."

Kate led me to a goat cheese stall, waving to several *fromagers* and *charcutiers* along the way. She bought a few tiny medallions of chèvre, no bigger than the palm-size, smooth river rocks I often collected from the Umpqua River, where my parents now lived, back home.

"These are small," she said, "but their flavor is so intense that you only need a sliver to feel satisfied."

We headed to the meat counter. Six or seven people stood in front of the meat case, and the six or seven men working behind the counter tended to each of them, looking full of pride in their red-and-black-pinstriped short-sleeved shirts, tailored to show off their muscular frames, and their white butcher's aprons, with just one strap that reached over one shoulder, so that the other arm was free to do all the movements required of a butcher.

The case was long, filled with so many cuts of meat—lamb, beef, pork, chicken, and duck, all laid out in such an advanced manner, organized by species but also by cut and by cooking method required—that stepping back from the case to take in its entirety resembled the experience of viewing a Chuck Close portrait up close and then from afar. I thought of the meat counters back home, where ground meat and sausages were king, with maybe a tied roast or two, and a lot of steaks and chops to fill the case out. This case held a good deal more cuts than that, many I didn't recognize.

Bright-red-and-white tile covered the wall behind the men, who spoke in polite French to their customers and ripped squares of butcher paper off of rolls hung from the ceiling. Dried sausages covered in white mold also hung from the ceiling. Through a glass window I could see hulking sides of beef and pork hanging—the pork looked almost as red as the beef, not the pale pink of the pork I ate back home. The counters they cut on were wooden, oiled by animal fat alone, with deeply worn grooves that had formed from years of cutting and cleaving. Even the sturdy red-and-white twine they used to tie the roasts was beautiful.

"This is a different kind of operation than the Chapolards have," Kate said. "These guys buy whole carcasses from small local farmers. They get them from the slaughterhouses and then cut them up and turn them into everything in this case. They're butchers only. But they are like a cheesemonger or an *affineur*. You won't find commodity meat at a shop like this in France."

Affineurs are the people who buy fresh cheese from cheesemakers and age them. And a cheesemonger is someone who doesn't just sell cheese, but handpicks it from the *fromagers*—the cheesemakers—and dairies. They are shepherds, in a way, of a product that requires careful curation, handling, and explanation. To think of a butcher this way felt new. They weren't just slingers of meat who opened up boxes of precut muscles and set them in their display case. They were trusted liaisons between the field and the plate. Their customers knew they had chosen the best product for them. Did we have anything resembling that back home in Portland? Not really. Not that I knew of. Not yet, anyway.

Kate looked at her watch. "Oh, no. Their train is already here!" It was time to go pick up her new students, Connie and Jenny, who wanted to learn all things baking during their stay.

"We'll eat heavy on the pastry this week," Kate said on the way to the train station. "Pie, croustade, tarts . . . *pâté en croûte*. We'll visit a few *boulangeries*." There would be salads made of fresh bitter greens, too, and meats both roasted and grilled, salted and cured, but the lessons would be in the language of flour, water, and butter. As Kate planned her lessons out loud, I wagered a guess that they'd never feel like lessons—they'd just feel like dinner.

SEVEN

We found Connie and Jenny standing out front of the train station, looking perturbed. Connie sported a well-coiffed, honeyed-blond bob resembling the rounded top of a brioche. Jenny's hairdo matched Connie's but had more of a bleach-blond hue. They hailed from Colorado Springs. When Connie spoke, I imagined the large bubbles that form on the sides of a glass of soda poured from the can, rising up into the air. Connie told us she'd homeschooled her children and that when they left home, her husband—who, she hinted, maybe had his own private jet—had bought her a commercial bread oven, which they'd installed in their three-car garage. She told us she wanted to turn an old house in her neighborhood into a bakery, "if only I could find a cheap Mexican boy to do all the work."

Jenny revealed that she'd just recently been pushed out of her family's medical supply company. She'd found God and dedicated her life to saving orphans in Brazil. And she was taking medication for her eczema, which made her nauseated and likely unable to eat most of the food we would be cooking for her. She seemed forlorn in a permanent way, whereas Connie seemed perpetually perky and positive. Connie was fit; Jenny was doughy. Connie liked to tell people what to do; Jenny wanted someone to tell her. They were perfect for each other. When I told them

I'd left my magazine career and wanted to become a butcher, Jenny blinked at me and then said, with hesitation, "How . . . interesting."

That evening, after they'd settled into their upstairs room with the Rapunzel-let-down-your-hair window, we got to work immediately, the four of us cooking dinner in Kate's kitchen, with Jonathan bobbing in and out as needed. While we cooked, Kate poured us little, delicate glasses of white Floc. It tasted a touch honeyed but with just the right amount of acidity—a perfectly placed semicolon between afternoon and evening.

Along with the Floc, she served us slices of cured duck prosciutto stuffed with foie gras. Eventually, Kate said, she'd take me to meet Jehanne Rignault, the duck-and-goose farmer who had made this epiphanic creation. But I wanted to know how she made it *now*—I'd never tasted anything like it.

"I'm still earning Jehanne's trust, so I don't have all the details yet," Kate admitted. "She salts two duck breasts with the skin on, then salts really fresh foie gras and puts that in between the breasts, and trusses it all together. She hangs it to dry for a few weeks in a cool, not-too-humid environment, like my piggery, until it's lost enough moisture that it's safe to eat, just like prosciutto or dried ham." I knew that all of the delicious, mostly imported, cured meats I bought at various delis back home were made this way, but I'd never really had to imagine the process, and it had never occurred to me to apply the process to duck breasts, let alone fattened duck liver.

The breast meat was a deep purple, the color of the venison and duck jerky I sometimes ate as a kid. It was firm but not hard, and it tasted salty like ham but with the added richness of that thick layer of duck fat and skin. The process of drying and fermenting had concentrated the meat and fat into a nearly inexplicable flavor—a deep, dark elementality. And in between these two already transformative duck breasts, yet another layer of complexity: the salted, semi-dried foie gras. Not a semicolon, but a colon, an opening onto an inordinate plane of complexity. Words to describe the flavor of that creamy, fattened liver loomed for an instant in my head, but just as I was about to utter each word, it disappeared.

"No wonder this shit is so controversial," I said. Anything with complexity always is, and the controversy always serves to reduce its object of ire to something so much less complex than the sum of its parts. Jenny blinked aggressively at me. She was doing that a lot. Maybe the God she had recently discovered didn't like swearing.

"They're trying to ban foie gras in California," Connie said, just a tad too righteously for my taste. I noticed she hadn't touched her slice.

"So are you planning on eating that?" I asked, picking a fight. Why would you come to Gascony and not eat this?

Kate chose the route of distraction. "I'm about to make a savory tart with tomatoes and chard and onions. Connie, why don't you make the crust."

Connie turned to find an apron. I grabbed the foie-gras-stuffed duck prosciutto from her plate and did not regret eating it.

THE NEXT DAY, Kate was hosting a group of journalists for lunch, so in the morning she led Connie and Jenny and me in the preparation of a *salade gascogne,* composed of crisp lettuces plucked straight from her garden and dressed lightly with a subtle mustard vinaigrette. We topped the salad with seared and sliced duck breast, foie gras pâté on toast, skewers of grilled duck tenderloin, grilled duck hearts that we marinated briefly in vinegar and shallots, and duck confit that Kate had made the previous winter and stored in tall jars in her piggery. This wasn't something people usually ate at home, Kate said. It was a dish for special occasions. The only thing I'd ever done with a duck as an adult was roast it whole and eat it in one sitting. When I was a kid, my dad and my grandpa had turned all the wild ducks they killed into jerky. In less than two hours, I'd learned how to turn one duck into five different dishes.

We set the table, poured a bottle of rosé into a decanter, and laid out the silverware and napkins and glasses just minutes before the journalists arrived. It'd been months since I talked to a journalist, so I chatted away with each of them, gossiping about the closing of several magazines in New York, the rise of Internet media, the controversial ousting of my old editor in chief at *Saveur*. They asked me if I would write about my time in France. I said I doubted it.

"I just want to be in the moment for now," I said to them, to which they offered me a knowing nod. I respected them for what they did, but I did not miss my old life.

THAT EVENING, after another late, long dinner—a savory meat pie, a tomato-and-corn salad, charcuterie, cheese—we all retired to our respective beds. Through the ceiling of my blue room, I could hear Connie and Jenny upstairs, gossiping, complaining.

"I can't believe Kate's making us do the dishes. How can she possibly charge us for this?" Connie said.

"Yeah, and why isn't she teaching us cassoulet?" Jenny said. "She's so free-form." Then, whispering, "Do you think she's a hippie?"

I laughed. Having grown up in Eugene, an epicenter for America's counterculture, Kate struck me as far from a hippie. Connie and Jenny had come looking for something specific. Maybe they wanted to wear toques and be yelled at for eight hours by a Gordon Ramsay look-alike with a stopwatch in his hand. Or maybe they just wanted to be waited on. But in the process of looking for whatever it was they hadn't found, they'd missed all the lessons Kate was throwing right onto the kitchen counter in front of them. This was Kate's magic. *Here's the difference between puff pastry and pie dough. Can you feel the difference with your hands? You're kneading that bread too much. Let it rest. See how it feels now? We don't make cassoulet in summer because the beans aren't ready, but tomatoes are, so let's make a tomato tart. We'll have to make a dough that can stand up to the liquid from the tomatoes. That stack of dishes over there—that's a sign we ate well tonight.*

"I don't know, but what about Skinny Girl?" Connie whispered back. They were talking about me. "What's with that butchery stuff? Do you think she's a dyke?" They both laughed.

When we're looking desperately for something particular—a culinary lesson, a person to love or hate, an easy justification for our bad behavior, a reason for why we don't fit in—or when we are guided solely by our self-serving expectations, we're willing to tell ourselves whatever story we need to, no matter whether the story is true or not, and, often, no matter the cost.

I had come to France in order to stop telling myself convenient but mostly untrue stories about myself. But I had also come to France to escape the stories I had no control over, the stories other people told themselves about me. I was naïve to think I could ever escape these.

A few days before I was fired from my magazine job, I'd knelt on my office floor next to the air vent that conveniently ran from my office to my editor in chief's, listening to her talk about me to our publisher.

"Completely insubordinate . . ."

"I can't have her around much longer . . ."

"The staff is scared of her . . ."

"Barely knows how to write . . ."

The story my editor in chief told herself about me wasn't altogether true—it merely justified my impending firing or layoff, or whatever it was, and, I suspect, her own insecurity. Nonetheless, I let her story shape my experience walking into the office every day. And until I arrived in France, I'd let it haunt me every day after I walked out of that office for the last time.

For Jenny and Connie, a wannabe butcher dyke and the lack of Mexicans to do their dishes had interrupted their expectations. A free-form hippie had swindled them out of their dreams and their money. They clearly weren't going to learn anything. They'd been duped. They were probably always being duped—by Mexicans and hippies and dykes and public schools and airlines.

On the other hand, I felt I'd been delivered to a French Shangri-la. I never wanted to leave. It wasn't that I minded so much being called a wannabe butcher dyke—I actually found it quite flattering—but I was determined to write my own story here, a truer one this time, whatever that might mean. Connie and Jenny and my editor in chief could go to hell.

Connie and Jenny left a day or two earlier than they had originally planned. We were all relieved.

EIGHT

Sunday supper at Dominique and Christiane Chapolard's house be-gan at one in the afternoon and ended about four hours later. When Kate and Jonathan and I entered their kitchen, they immediately put down what they were working on, kissed me on each of my cheeks, and hugged me as though we had known one another for years. *Bonjour! Ça va? Bonjour! Ça va?*

Delighted is not a word I use lightly, ever. But it was, I think, how we all looked and felt in that moment. And this despite the fact that these people didn't know me, despite the fact that the payoff for taking me under their wing—I was unable to pay them much to do so—was prob-ably not very clear. They were taking a chance on me, an unknown entity with no butchery experience, yet they didn't seem the least bit worried. Instead they were hugging me, practically singing in French to me, pushing paper-thin slices of *jambon* into my hand. I wanted to stand there forever with them, absorbing their warmth.

As they prepared the meal for their guests, conversing enthusiasti-cally with Kate, who only occasionally translated for me what they were saying, I studied the two of them. They both looked to be in their early fifties. Dominique had curly brown-and-gray hair and what struck me as a quintessentially French mustache—thick, salt-and-pepper, with ends that naturally curled up—accentuating his kind smile. "He looks even more French when he wears his beret at the market," Kate told me. He was shorter than me, stocky, solid, firmly connected to the ground. The considered, kind meter of his voice calmed me.

Dominique's wife, Christiane, stood a little shorter than him. Her voice often lilted up into a high, singsongy, operatic register. She possessed a stern, hawklike attention to the goings-on around her, but she also appeared friendly, open. I immediately liked my new French *mère* and *père*. After a few minutes, they shooed us outside to join the other guests on the lawn.

At least twenty people milled about out front of Dominique and Christiane's house. The other Chapolard brothers, Bruno, Marc, and Jacques, each greeted me in a slightly formal manner, with kisses on my cheeks but no hugs, and introduced me to their wives, and at least a few of each of their children, some fully grown, others well on their way to adulthood. Madame and Monsieur Chapolard, François and Antoine, the brothers' parents, kissed me on my cheeks and nodded silently. Kate's svelte, tan handyman arrived with his girlfriend and child. A few of the Chapolards' fellow *producteurs* at the outdoor markets where they sold their pork each week showed up, too.

"Oh, there's Jehanne!" Kate said, pointing to a beat-up, dust-covered farm truck coming up the driveway. "She's the duck woman I want you to meet."

I had not stopped thinking about Jehanne's foie-gras-stuffed duck prosciutto.

"We need to take you to her *ferme auberge*." Kate was talking about Jehanne's farm, where Jehanne fed overnight guests lunch and dinner and where she raised, slaughtered, butchered, and processed a small number of fattened ducks and geese per year and turned them into various forms of charcuterie, which she sold at markets, just like the Chapolards.

Jehanne had a weathered, tan face, graying blond hair, a square jaw, big brown eyes, and a long nose and forehead. She was aging, tired, and gorgeous all at once. As Kate explained to her why I had come to Gascony, she looked me up and down with squinting, scrutinizing eyes. *She wants to be a butcher. She wants to learn about meat. Can she come watch your next slaughter? Can you show her how you butcher the meat while it's still warm? How you make your foie gras taste so good?*

"That is a secret," Jehanne said in English, not smiling, but with a wink to me. She seemed reserved and pragmatic, but I sensed that she harbored a wilder past. We made arrangements for a visit.

Christiane offered us each small glasses of hard apple cider mixed with a little pear juice and told us to find a seat at the long table they'd set up outside. We first dined on melon wrapped with the Chapolards' salted, smoked, and dried *jambon*. Then Dominique brought out a heavy ceramic pot filled with one of their pork shoulders, which he had stuffed with Gascon prunes and roasted, along with onions and golden potatoes that had caramelized in the oven, turning them creamy and sweet. Just as that steak Kate had bought me in the train station tasted more like meat than any of the meat I ate back in the States, so, too, did this pork. I told Kate this, and she assured me that once I started working with the Chapolards, I'd understand why.

We switched to a red wine from Bruilhois, a wine whose color was nearly black and whose flavors reminded me of ingredients of the same hue. Blackberries. Black currants. The color of that cured duck breast, which is now the color I think of when I think of Gascony.

Every time Kate and I bought wine at the grocery store or went to the Buzet wine cooperative and filled our big plastic jugs with wine, I marveled at how low the prices were, given the quality. Sure, some bottles were more expensive, and perhaps more nuanced than others, but none of it was swill, and "cheap" didn't mean bad. Nor was "expensive" ever as expensive as it was in the States. Kate told me that the French thought of wine as a native right, an essential ingredient that everyone should be able to afford. Also, as with the way they ate meat and cheese, I noticed people didn't drink very much of it. The modest pours in our glasses were for tasting and savoring, not for gulping or getting drunk, which, given my drinking habits of late, I found a bit challenging at first.

Gascon portions of food *and* drink were generally modest compared with American portions. And yet Gascons (and maybe the French in general) seemed to spend more time eating than Americans do—even when my goal was to have a long, leisurely dinner party for friends, they never lasted this long. The food on their plates also represented quality, not quantity, and they did very little snacking in between. In America, we take as little time as possible to finish our food, even though we have more of it on our plates. We also seem to be eating all the time—we are so very hungry, but never sated.

I read a study recently that asked people in France and the United States which word came to mind when they thought of heavy cream: *whipped* or *unhealthy*. An overwhelming majority of Americans related the word *unhealthy* to heavy cream. The French imagined the cream whipped, frothy, and delicious above all else. When asked whether they eat a healthy diet, a majority of Americans answered no. A majority of French subjects said they did eat healthy diets—whipped cream, duck fat, foie gras, and all. Moderation, it seemed to me, was key here, but moderation made sense only because the quality of the food and drink was so high.

Kate once complained to me that when she took her American students to eat at the Chapolards' home or those of other French friends, and the hosts offered her students a second helping of food, her students often replied, "No, thank you. I'm stuffed!" or, "I couldn't possibly eat another bite," while pointing at their bellies.

"In America, that's a compliment, but in France it's rude to say that," Kate said. "Now I teach my students to say, '*J'ai bien mangé, merci.*'" I have eaten well, thank you.

After supper, everyone sat back in their chairs and began talking a little slower. Dominique brought us small cups of coffee, which somehow made me sleepier. To wake us up and to cleanse our palates, Christiane served us all cold shot glasses of a slushy, frozen strawberry liquid, a cross between sorbet and a snow cone. We sang "Bon Anniversaire" to Grand-Père. Everyone spoke in French. Kate had long ago stopped translating for me. I felt sated but not stuffed, sleepy from all the words I could not understand. My eyelids drooped, and my brain denied me all possibility of linguistic comprehension. I noticed that Grand-Père had dozed off and Dominique's eyelids were fluttering. But we all stayed at the table.

AT FIVE, Kate drove Jonathan and me back to her house. We each found a hammock under the tall fir trees by Kate's boules court and swung silently in our nets. It was humid and sticky out, and the heat pushed us into a deep, warm, muddy, post-Sunday-supper coma.

Before I fell asleep, I remembered an e-mail I'd received from Will that morning, hinting that he was worried I was having an affair over in France. I hadn't been very good at writing or calling, so I could understand why he wondered.

I am having an affair with myself, I thought. *And I quite like it.*

When I awoke, the sun was setting, and Kate stood over me. She'd painted on bright-red lipstick, put on a pair of midnight-blue glass earrings, and covered her shoulders with a paisley shawl in fuchsia and purple. She'd showered and found her second wind. How did she do it?

"Ready to go to the *fête de l'escargot*?"

The snail fest in the main village of Sainte-Colombe-en-Bruilhois. I'd been looking forward to it all week.

"I'm ready," Jonathan declared with a level of enthusiasm I'd not yet heard him express. He rubbed his hands through his messy hair, in an attempt to smooth it down. Perhaps, I remember thinking, like me, he was slowly letting down his guard, learning to relax and enjoy himself, after years spent working way too hard and taking everything way too seriously.

"I'm ready, too," I said, groggy, with one eye open. I ran up to my blue room to put on some pants and a sweater—it still grew cold at night in this river basin.

At the fête, one of many we would go to during my time in Gascony, hairy, shirtless, sweating men stood over open flames, stirring huge paella pans of escargot. Kate's American friends Alvin and Renee saved a table for us. We poured a little more wine, scooped garlicky snails out of their shells with flimsy toothpicks, and dipped our fries in mayonnaise. A gypsy band played from a small stage. Wispy clouds scooted lazily along the horizon. My kind of affair.

THAT NIGHT, lying on my bed in the blue room, a French word settled into my head and prevented me from sleeping. I wasn't sure what prompted its arrival, but I silently repeated it over and over. *Débrouillard*. Perhaps I'd heard someone say it in passing, though I could not remember the context. I turned my bedside light back on and looked up the root word, *brouillard*, in the little French dictionary I carried everywhere. It meant "fog." *Débrouillard*, I guessed, must mean "to defog."

My first week in France had been all about the fog of travel and jet lag, the fog of being a new person in a new place. By the end of the first week, I'd just begun the process of defogging.

But when I looked up *débrouillard* in the same dictionary, it gave the meaning as "resourceful." I liked the notion that to find one's way out of the fog required resourcefulness. And then I remembered how I'd first become aware of the word. It was a few years back, when I'd been preparing for an interview with Anthony Bourdain while he was in Portland to film an episode for one of his travel shows and come across mention of the word in his book *The Nasty Bits*. Bourdain talked about the French word in relation to the term *System D*, a shorthand used in kitchens to describe a manner of thinking fast on one's feet in response to a spur-of-the-moment challenge that arises. Employing System D, in Bourdain's world, meant improvising to get an urgent job done, solving complex problems with whatever resources you had to work with, even if they were few, like fixing a broken piece of kitchen equipment with a teaspoon or turning a frozen mini-pizza, frostbitten from too much time spent hiding in the recesses of the freezer, into something presentable when you've run out of fancier hors d'oeuvres for a corporate cocktail party.

"Every kitchen has one evil genius who's tolerated," Bourdain once said in an interview with the *Harvard Business Journal*, "someone you turn to when all else fails—a rule breaker, a scamp who's willing to make a hard and sometimes unlovely decision for expediency. There's actually a name for this person—the *débrouillard*, the person who gets you out of a jam."

The *D* in *System D* might be traced back to the noun *débrouillard,* or it might refer back to the verbs *se débrouiller* and *se démerder,* both of which mean "to make do, to manage, especially in an adverse situation." It seemed to me that *débrouillard* wasn't just a noun or an adjective or a verb but a philosophical state of mind, one that required you to be fully in the world as it presented itself to you.

I couldn't truthfully say that I'd consciously chosen this particular philosophical state of mind—unemployment and an early onset midlife crisis surrounding the meaning of love and the value of work had forced me in this direction. Everything had gone to shit back home, so I'd become my own *débrouillard* and made a risky and perhaps unlovely decision to leave everyone and everything I knew. In the spirit of *débrouillardise,* I was attempting to forge a clear path for myself into a world dominated by men, a world that is often associated with the darker side of eating, what with the blood and guts and bone involved in turning animals into dinner. The next morning, when I stepped into the Chapolards' cutting room for the first time, I was forced, finally, into the immediate present, into whatever stood right in front of me, forced to draw from my own resources in order to make my way. And in making my way, I felt tremendous relief.

NINE

We were all dressed in white. White rubber boots. White coats. The Chapolard brothers wore coats with hoods, which they pulled over their heads to keep warm and to keep their hair from contacting the meat. Dominique, Marc, and Bruno all held knives in their hands. Jacques roamed around outside, tending to the pigs in his blue coveralls. Every once in a while I caught a glimpse of him through the one small window in the Chapolards' *salle de découpe*, the cutting room, walking from the pig barns to a large silo of grain to a tractor and back again.

All five women in the room were gathered around one table, wearing white hairnets, hard at work. Dominique's wife, Christiane, stood next to Marc's wife, Cecile, who stood next to Marjorie, a young Frenchwoman who was enrolled in a nearby agricultural school to become a butcher and had recently begun staging with the Chapolards. Kate was there, too, to help orient me on my first day, to translate for me, but soon my French mother hen, my linguistic lifeline, would leave me and I would have to navigate this language on my own, in this tiny room with so many people and so many cuts of meat I did not know.

The cutting room was kept at around forty-five degrees, but outside it was summer, so the brothers were dressed in short white shorts, exposing their hairy legs, nearly as thick as whole hams, to the air-conditioned elements. They didn't seem to mind the cold. I, on the other hand, had covered my skinny legs in tight black jeans. A black turtle-neck and a cotton sweatshirt kept my torso and neck warm, but barely. I was swimming in a white cotton butcher's coat meant for someone much larger than me, and I'd buttoned it up to my chin and turned the collar up. The Chapolards were a tidy clan, but nevertheless, the floor was slippery with fat and water—par for the course in the land of butchery. Even when I tried to stand still, I slid in my too big boots, a clumsy fawn just learning to use her legs and feet.

The unspoken rules of the cutting room: Keep your elbows in. When one person changes positions, or needs salt for the *jambons,* or is ready to bring a lug of meat to the grinder, adjust accordingly. It was that tight. The Chapolards' *salle de découpe* was the size of a deluxe Airstream at best. Part of an old aboveground winemaking cellar built before the brothers' grandfather and great-uncle bought it after the First World War it was, like Kate's *pigeonnier,* made of limestone rubble, rock, brick, and mortar. Worn brown and gray stone outside. White vinyl walls and fluorescent lights inside.

The long, narrow cutting room had four successive doorways on the right-hand side, leading into smaller rooms. Dominique began giving me a tour. First, on the right, was a large walk-in refrigerator, even colder than the room they cut meat in, with an entrance from the outside so that they could pull up with their truck and unload carcasses easily. The walk-in was mostly empty this early Monday afternoon, save for a couple of buckets and sets of organs from each of the ten pigs that were killed at the abattoir that morning, the abattoir where I had earlier—very early, in fact—witnessed the slaughter of that seven-hundred-pound sow, owned by a different farmer, who now had a whole lot of sausage to make. Dominique pointed to a blue bucket full of bright-red liquid.

"*Le sang,*" he said. Blood.

"*Le sang,*" I repeated.

He pointed to a white bucket full of a lacy netting of caul fat, the thin membrane that surrounds the digestive organs.

"*Crépine,*" Dominique said.

"*Cray-PEEN,*" I said back.

"And that's pig intestine," Kate said, pointing to another bucket. "They call the intestines and the caul fat white offal. And this is the red offal," she explained, pointing to the sets of organs hanging from the ceiling. "The lungs, heart, and liver are left attached to the trachea and pulled out of the carcass all in one piece so they're easy to hang and store until they're ready to work with them." Each set of red offal looked like a Dalí painting, spilling off the tip of an S-shaped metal hook. I recognized only the heart.

IN FRENCH AND ENGLISH, Kate and Dominique caught me up on where the Chapolards were in their weekly cycle.

It was just after lunch on Monday. On Sunday, before Grand-Père's birthday party, the Chapolards had dropped their ten live pigs off at the abattoir to settle into their new surroundings—a barn attached to the abattoir—the day before slaughter. Earlier this morning, after the slaughter, Marc had retrieved the heads, the red and white offal, and the blood to be used for *boudin noir* and brought them back to the *salle de découpe*. They'd spent the morning processing all of this.

"This afternoon, Jacques is going to the abattoir to pick up the carcasses," Kate told me. "And you're going with him."

Eight to ten pigs split in half every week. That was somewhere between four hundred and five hundred pigs a year. It sounded like a lot to me, but what did I know? I asked Kate whether it was.

"It's not," she said, "not compared to factory farms. This is a very small pig farm."

I asked her if she knew how many pigs are raised and slaughtered by factory farms each year.

"A lot," she said. "This is a different kind of operation." By comparison, a confined animal-feeding operation in America can raise upward of 150,000 per year.

WHILE WE WAITED for Jacques to arrive, Kate and Dominique tried to give me a better sense of scale. By the end of each week, they explained, after three different outdoor markets, the Chapolards had usually sold every part of every one of their ten weekly animals to their customers, save for the bones, which they composted for use on their farm, and whatever they took home to feed their own families. When a customer bought a slice of ham from them, the Chapolards could vouch for every part of the process that transformed one of their pigs into that slice of ham. They grew the grain to feed their pigs. They raised the pigs themselves. They owned their cooperative abattoir with other small farmers. They did all the cutting and curing. They sold the meat at outdoor markets. They owned every part of the process, and *this* was their appeal. By French standards, they lived modestly, though comfortably—a kind of modern-day middle-class peasant.

Though there were still plenty of small meat producers in France, Kate told me the Chapolard model was increasingly rare. Most meat producers raising animals in numbers similar to the Chapolards sold them to small local butcher shops like the one Kate had taken me to in Agen, which then took care of the rest of the processing. Much bigger meat producers, of course, sold their animals at auction or contracted with larger meat distributors and conglomerates, which stocked the meat counters of most grocery stores. France had its share of factory farms just like America. But France also still had farmers like the Chapolards and, to my knowledge, America did not. Most small farmers I'd met didn't grow their own grain and didn't butcher their own animals, let alone make pig-head pâté and blood sausage to sell at farmers' markets.

Four major conglomerates process roughly two-thirds of the one hundred million hogs raised in America each year. Technically, these conglomerates, like the Chapolards, own all of the animals that become meat, although these companies typically contract with farms— confined animal-feeding operations—to raise them. They also, typically, own the slaughterhouses and processing facilities that turn those animals into food. But these conglomerates differ from the Chapolards in one major way: division of labor is king. By the time a steak from one of these operations gets to your table, it may have passed through dozens of hands and crossed many state lines, and whoever is selling it to you most likely can't tell you much, if anything, about how it got there—they're not even legally required to know.

This division of labor adds up to division of knowledge. I'd found this out when I wrote that story about *bavette*, back when I still thought all butcher shops employed people who knew how to cut up whole carcasses, who knew which part of the animal each cut came from—back when I thought chefs knew this stuff, too. In fact, very few of the people serving or selling meat in America, let alone the people buying it, knew much about where it came from.

I WANTED TO SHARE my *bavette* story with Dominique, to tell him what it was like to be a curious person asking questions about meat in the States, but Dominique and Christiane motioned for Kate and me to follow them to another room at the very back of the *salle de découpe*. When we opened the door, a steamy, unctuous warmth enveloped us, a welcome relief from the chilly atmosphere of the cutting room. My nose immediately filled with the strong scent of meat and bones cooking with bay leaves, leeks, onions, and carrots in a vat of water that was wider than me by several feet and almost as tall. The ingredients, Kate told me, were slowly working their way toward a rich, gelatinous, flavorful symbiosis.

In the vat, Christiane said, they'd placed the equivalent of six split pig skulls, with the cheeks and other bits of meat, fat, and skin all still left on them, along with several tongues and a few hocks and trotters. This would cook for six hours, and when it was done they'd pick the slow-cooked meat from the bone, gather the very tender tongues, and combine it all with some of that gelatinous liquid and fat in a heavy terrine to make *pâté de tête*, otherwise known as *fromage de tête*, or head cheese in English.

A chef back in Oregon once told me that if I saw "pulled pork" on his menu, and it had quotation marks around it, it was actually head cheese. Why did he have to disguise it in this way? Because his customers, he explained to me, would never buy something called head cheese, and enough people knew French to be able to translate *fromage de tête*.

I told Kate this, and she told Christiane and Dominique.

"Why don't Americans like head cheese?" Dominique asked, confused.

"Because they don't know what it is. They think it's brains and eyeballs and everything else," I said.

"And what is wrong with brains and everything else?" Dominique asked, shaking his head.

"Because we don't like to be reminded where meat comes from," I said. We don't like brains because brains remind us that the meat we're eating came from an animal like us. We don't like eyeballs because they can stare back at us.

Kate translated, and Dominique stared at me for a long while, looking exasperated and a touch forlorn. "Why would you raise an animal for food and then not eat every part?"

I did not have a good answer for him.

I no longer remember what I envisioned when I first heard the term "head cheese," but I know now that head cheese comprises some of the most flavorful pieces of meat from the entire pig, and not brains or eyeballs. And yet, the chef back in Portland had told me, even though he called it "pulled pork," it was a hard sell once it arrived on the plate.

"Too foreign-looking for kids like us," he'd joked. *Kids like us.*

BACK OUT IN the cold cutting room, Dominique told me that once Jacques and I returned with the carcasses, they'd begin to work on those, creating piles for each kind of cooked or cured product they'd make that afternoon and into the next day. Tuesdays they continue with this work, stuffing sausages, cutting and trimming chops and roasts, preparing brochettes (skewers), and trimming the belly for *ventrèche*, which, Kate explained to me, was the Chapolards' version of American bacon—basically, salted and smoked belly that is liberally flavored with black pepper, but without the sugar that American bacon cure usually includes. All this in preparation for the first two markets of their week, in Casteljaloux on Tuesday and Lavardac on Wednesday. By the end of the week they'd have enough fresh cuts and cooked charcuterie like head cheese and other pâtés, plus sausages, hams, and bellies that have been salted and smoked or hung to dry in previous weeks, to sell at two more markets on Saturday, in Nérac and back at Casteljaloux. The brothers and their wives took turns working the markets, sometimes even bringing their children to help them, such that someone was always available to fulfill the constant production needs in the cutting room and on the farm.

The Chapolards faced a dizzying amount of work each week, and I was about to ask if they ever felt they didn't have enough people to get it all done when Jacques entered. He was still grinning, just as he had been when we greeted him upon our arrival at the Chapolards' farm, right before he'd run home for lunch. Kate explained that he wanted to give us a tour of the farm before he and I drove to the abattoir to pick up the carcasses, so we stepped out of our cutting-room attire and into the sunny outdoors, squinting.

Out back of the cutting room, we stood in an unkempt courtyard of dirt and scrappy weeds surrounded by barns and other outbuildings. Old equipment and piles of materials—wire netting, rotting boards, rolls of gauzy material—were strewn everywhere. Grass grew in between the springs and gears of some of the old equipment.

"You know it's a working farm when it looks like this," Kate said. "If it looks clean and tidy, the owners probably have a day job."

A short grain silo stood at attention next to the open-air barn that housed the Chapolards' older pigs that were headed for slaughter. Jacques explained that the silo was where they stored the grains that they fed their pigs, grains they grew themselves on the property—corn, wheat, barley, oats, sunflower seeds, and féverole, a type of field bean.

I asked Kate why they didn't pasture their pigs, a phrase I'd heard chefs and farmers toss around back home. She explained that "pasturing" in the context of pigs meant something different than with cattle or lamb, both ruminant species, meaning the multichambered structure of their stomach allows them to efficiently digest fiber such that they can subsist and fatten up with a diet of grass alone. Pigs are a monogastric species, she explained, meaning that they are inefficient digesters of fibers like grass such that, although they do love rooting in and eating pasture, they are unable to survive, let alone develop into edible, delicious meat, on pasture alone. They need a balanced ration of grains that have been processed—cracked, rolled, or soaked—so that their stomachs can efficiently digest them. "Pasturing" a pig is really about giving them exercise, and not so much about their diet. The Chapolards didn't have a lot of acreage to work with—only about a hundred hectares, the average size of small family farms in the area—and since they'd chosen to grow their own pig feed, their small parcel of land couldn't accommodate both grain and pigs.

From a smaller, concrete-and-brick structure, a cacophony of squeals and grunts spilled out toward us. It was a long, low, somewhat narrow building with a tin roof and just a few open portions cut into the walls for airflow. At the front door, Jacques explained that this one held all of their nursing mothers and babies, all of whom were easily spooked. The full walls and minimal number of windows in this nursery ensured that distractions were kept to a minimum and that the new piglets stayed warm.

Jacques slid open the wood-and-metal door, instructing us to be quiet and move slowly as we walked in. The room smelled like pulsing, breathing, new, hungry, dirty, shitting life, a scent I hadn't been privy to on a regular basis. Kate explained that all of the manure gets put back out into the fields, where it decomposes and renews the soil with nutrients that make their grain grow.

Although it was relatively small, the darkness of the interior made the space feel cavernous, as if we were standing inside the womb of a twelve-thousand-pound mama. The sounds of the babies and mothers pinged and echoed off the walls. I watched as eight or so tiny piglets violently shoved their mouths into the teats of their mama pig, who was lying on her side, surrounded by a loose, metal crate or cage, whose "ceiling" and "walls" resembled those metal farm gates you sometimes see at the entrance to a road leading into an animal pasture. Jacques told us that the structure was used to keep the sow from rolling over and killing her babies. Later, I'd find out the name for this practice: farrowing. Many farmers struggle with whether it is humane to sacrifice the mother's freedom to move around in order to save the piglets from possible death. Some of them believe that, so long as you can make a mother pig feel safe—with walls made of hay bales, for example—as opposed to trapping her in what amounts to a tight metal cage, she won't roll onto her babies. But on the Chapolards' farm, Jacques talked about farrowing pens very matter-of-factly. *This is how it is done.*

As we walked farther into the barn, the piglets grew louder, so Jacques, looking a little nervous, motioned for us to leave. "He doesn't want us to stress the mamas and babies out," Kate explained.

Jacques took us across the dirt courtyard to another barn, where all of the pigs destined for the dinner table resided once they'd moved beyond nursing to solid food. The barn was open to the elements, with gauzy netting hung from the rafters around the perimeter to keep out dust and bugs. The pigs weren't cramped in there—they had plenty of room to move around—but they weren't free to roam the farm, either. Jacques explained that, in total, they had three barns for "grower" pigs, those that would soon be sent to the abattoir, plus one for sows, one that was a nursery for new piglets and mothers, and one for "weaners," who had stopped nursing and were transitioning to solid foods but were not quite ready to live in the grower barns with older pigs.

Jacques nodded his head, signaling that our tour was over.

"Okay," Kate said. "He's going to take you with him to pick up the carcasses, and then you'll come back here and work some more. This is my cue to leave. Have fun." I'd be working with the Chapolards for the rest of the afternoon, then spending the night at Dominique and Christiane's house, in a nearby village. I looked at Kate, feeling a bit desperate. I was the new kid, and she was dropping me off for my first day of school.

She hugged me.

"What am I going to do without my translator?" I asked.

"The only way to learn French is to get me out of the picture. You'll do fine. I'll come back tomorrow." And with that, she settled into her car and drove away.

I turned to Jacques, who was *still* grinning as he motioned for me to follow him to a tall white van. As we drove down the long dirt driveway, I listened to the rattling sounds of our empty metal driving box and wondered what it would sound like with twenty sides of pig in it.

Jacques asked me if I liked the Beatles and then started singing "Back in the U.S.S.R.," urging me to sing along, but I couldn't remember any of the words.

TEN

I'd already been to the abattoir, a small complex of tall white buildings with windows at the top of each wall, in the nearby town of Condom, earlier that morning to witness the slaughter with Marc and Kate, but this time we entered through a different door. We parked our van at a set of sliding metal garage doors—the loading dock, clearly, as there were other men with vans, too.

We slid open the garage door we'd parked in front of and then entered an echoey, dark concrete corridor with a few inadequate fluorescent lights strung high above us. We took a right, walked down another corridor, then pulled open a heavy insulated metal door, which led us into a cold antechamber. Jacques instructed me to dip the soles of my shoes in a shallow, soapy tray of water on the floor, then we pushed open another, heavier door, which triggered a burst of cold air. I was lost almost immediately—and this, Kate had told me earlier that morning, constituted a *small* slaughterhouse.

We ended up in a vast, soaring refrigerated chamber full of hanging half and whole animal carcasses that had been killed that morning. Most were pigs, but a few were lambs, or maybe goats—it was difficult to tell with their skin off. One or two carcasses looked a lot bigger, maybe beef, maybe something else, but they were also skinned, so I could not be sure. All I could see was a series of gargantuan muscles, long white tendons. How different the pig carcasses looked from the other carcasses—with so much more fat, and all their skin still wrapped around them.

Each of the carcasses hung by one or two of its back legs—depending on whether it had been split in half or left whole—by way of thick metal S-hooks attached to a system of ceiling rails high above me that ran out the doors into other rooms. When I say they were hanging by their legs, I mean that someone had taken a knife and cut a hole between the Achilles tendon and the anklebone. The hook was inserted into that hole, which wouldn't rip because it sat between that very strong tendon and bone.

Jacques stopped to talk to a short man in white coveralls, so I began counting the carcasses. About seventy halves of pig, so thirty-five or so whole pigs. Maybe ten lambs, or what I assumed were lambs. And just a few of the larger beeflike animals. Could they be horses? I wondered. Kate had told me that it was normal for people to slaughter horses here when they got too old for other agricultural uses, and turn them into food.

I stared at a rack of shelves with hooks welded every foot or so along the front of each shelf. A skinned lamb head hung on each of the hooks. The eyes looked out, blank.

I asked, in stuttering French and English, how much each side of pork weighed, and Jacques told me, by way of hands and fingers, that each half weighed about two hundred pounds.

Then Jacques demonstrated for me how we would move the Chapolards' twenty sides of pork into the van. He walked over to one of the hanging sides and stood in front of it. What does a side of pig look like? Imagine a pig cut in half along the spine. Each side has one half of its rib cage, one half of the belly, one half of the spine, one back leg, and one front leg. With both of his thick, meaty hands, he pushed a side forward along the system of rails above us, his upper body pressing forward toward the door, while his legs pressed back toward the concrete floor. He appeared to do this with little effort. I grabbed on to another side just as he had shown me, but the strength of my hands felt inadequate for pushing this much weight. In an attempt to conjure more force, I awkwardly bear-hugged the carcass, with my neck and face cranked around the side of the pig so I wasn't staring into its inner cavity and I could see where I was going, and walked the carcass along this way, although at times it seemed as though the carcass was walking me. The carcass was, as Dominique had cautioned, still slightly warm. Hugging this animal body, my own temperature rose slightly. It was difficult not to feel a visceral overlap at the liminal place where my skin touched the pig's skin. Jacques turned around to check on me and laughed. It was a friendly laugh, but he clearly thought my technique wasn't the best way to get the job done.

We moved all twenty of the sides this way, lining them up on the ceiling rail next to the door where we'd parked our van.

"*Très bon!*" Jacques declared every time I arrived with a carcass. Sometimes he even clapped. He never stopped grinning. Jacques's coveralls were spotless. I looked down to discover my black turtleneck smeared with white fat and stray bits of meat.

When we were ready to load the carcasses into the van, Jacques grabbed each one with a mechanical hand that hooked onto their feet and guided them gently into the truck, where he hung them again by their back legs on hooks dangling from the ceiling.

Jacques motioned for me to stay put and then disappeared into the labyrinthine halls of the abattoir. A few other men were in the process of loading carcasses into their own vans. They all looked at me with faces I could not read. When I caught one of the men's eyes, he nodded at me, then looked away. I was a new face and, I assumed, not an expected one. I got the sense that the Chapolard wives never picked up the carcasses.

I picked the pieces of fat and meat from my shirt and flicked them to the ground. I'd just bear-hugged my first pig carcass—or, rather, ten halves of pig carcasses. Pig carcasses that, somehow, we would turn into dinner. Into many dinners. When people sat down to eat them, what would they know about the pork in front of them? Would they know that an American girl had hugged their dinner on her first day in a Gascon abattoir?

ELEVEN

After Jacques, Dominique, and Bruno transferred the pig halves into the *salle de découpe*'s walk-in, we had no time to rest. Jacques had to get back to his snorting, pooping, hungry pigs, and Bruno, Cecile, Marjorie, Christiane, and Dominique needed to get to work right away, cutting, trimming, and sorting carcasses, putting some meat in a pile for *saucisson*, and other meat in a pile for *saucisse sèche*.

"*Quelle es la différence?*" I asked Dominique, in stilted French.

This question sat at the center of everything I learned in Gascony. What was the difference between *saucisson* and *saucisse sèche*? Between pain and suffering and stress and discomfort? Between life in a pasture and life in a barn?

Dominique answered, but my grasp of French verbs and conjugations, as well as my grasp of meat, was too basic for me to be able to understand him. Luckily, Dominique was a patient teacher and a creative one, so he took me to one of the rooms off the main cutting room and closed the door behind us.

"Room" is actually a terrible way to describe it. It was more like a primordial meat cave with a few modern flourishes thrown in. While cavemen weren't in the habit of hanging *saucisson*, the smells of that space connected me to a former way of life that I desperately longed for, without knowing why.

All manner of nearly dried or drying sausages hung from two dozen horizontal wooden dowels that the Chapolards had attached to a vertical metal structure. Americans are more familiar with the Italian word for these sausages: *salami*. Dominique schooled me in the French version.

"*Saucisse sèche,*" Dominique said, pointing to a somewhat shriveled, purple-looking sausage, more like the color of Jehanne's salted and dried duck breasts. Flecked with ground meat and what looked like fat, the *saucisse sèche* was about an inch in diameter.

"*Saucisson,*" Dominique said, pointing to a much larger dried sausage the color of the tiled roofs in Toulouse. I knew that the word *saucisse,* when used alone, referred to fresh sausage, so I deduced that *saucisse sèche* meant dried sausage and that *saucisson* was just a larger version of *saucisse sèche*. Dominique gestured for me to squeeze the *saucisse sèche* and the *saucisson* between my thumb and forefinger. Each had a little give to them, but the *saucisson* seemed softer than the *saucisse sèche*. They were all clearly made out of meat and fat and encased in pig intestine, so why did they look so different? "Could it be they came from different parts of the animal?" I tried to ask Dominique in French. He pointed to his butt and thigh and said, "*Jambon,*" followed by a lot of other French words I didn't understand.

Later, Kate would explain that *saucisson* is made from lean meat from the hind leg, or ham, that has been completely trimmed of fat, sinew, and connective tissue, with 20 percent of fatback added back in. *Saucisse sèche* is in some ways a by-product of *saucisson,* made from meat from other parts of the animal plus a good amount of the fat, sinew, and connective tissue that was trimmed in the process of making *saucisson*.

Only when I finally understood this did I come to appreciate the deep level of resourcefulness that the Chapolards employed when it came to using the whole animal.

Dominique took me back out to the cutting-room tables, where Bruno, Cecile, Christiane, and Marjorie were hard at work. I watched him thinly slice a long cylinder of fresh, lean muscle, for which I had no name yet.

"For Christiane's paupiettes," he said, making me repeat the word, *Pope-ee-YETS*. He handed the slices to Christiane, who wrapped small scoops of ground pork inside of each, then a thin slice of what looked like bacon around that.

On another table, Marjorie rolled three-inch-wide pieces of skin into tidy little packages and tied them with red-and-white butcher's twine. They would sell these at this week's market, too.

Nearby, Cecile formed fist-size balls of what looked like ground meat and wrapped them in caul fat.

After slicing meat for Christiane, Dominique began trimming a flat piece of meat that looked like what Christiane had wrapped around the outside of her paupiettes. He pointed to his belly and said, *"Ventre."* Then he walked into another room and returned with a beautiful, thick, tied roll of that same piece of meat, covered in ground black pepper. *"Ventrèche,"* he said, encouraging me to smell it. It smelled like black pepper, wood smoke, and salted meat funk, the good kind. "French bacon," he said in English, pronouncing it *bay-cone* and letting out a big, happy sigh afterward.

Marjorie began cutting cubes of meat from a large muscle for brochettes. The afternoon moved on like this. *What is this?* I asked. *How do you say that? We'll use the liver for this pâté. We'll use this fat and skin for gratton. Soon it will be time to salt these jambons.* Christiane demonstrated for me how to put the chunks of meat Marjorie was cutting onto wooden skewers, making sure that the grain of the muscle of each piece ran in the same direction on the stick. Everything felt chaotic and new to me, but underneath the chaos I sensed a well-thought-out order, a rhythm and philosophy that only the Chapolards knew.

MARC, the one brother who did not look like the others—tall, darker-skinned, with a longer face and thicker lips, the only Chapolard without a beard or a mustache—walked in and nodded at me courteously, but did not kiss me on my cheeks or smile. *Le sang,* I heard him say to everyone in the room. It was time to make *boudin noir.* Blood sausage.

Marc motioned for me to follow him into another small refrigerated room off the main room, filled with shelves of product that I could not yet identify. There were metal molds of pressed meat. A few more of those blue plastic buckets of cleaned intestines. Another bucket of beautiful, lacy caul fat, which, in such a large amount and floating in water, took on the appearance of a deep-sea mutant, like a cross between a jellyfish, an octopus, and a manta ray. A machine with a very short conveyor belt and two metal rollers took up one corner. It wasn't industrial-size, but it was big enough that I felt I should proceed with caution. I pointed to the machine and asked what it was.

Marc grabbed a long piece of pigskin with a thick inch of fat still clinging to it and sent it through the two metal rollers to show me how it cleanly separated fat from skin, handing me a piece of each to feel and smell. The skin was smooth, a pale peachy pink, and full of gelatin, the gelatin that binds meat together in their *pâté de tête,* the gelatin that thickens soups and stews and cassoulets. The fat was creamy and white, and just a touch sweet-smelling.

How often does one get to do this? To hold this raw skin and fat and meat and bone in one's hands, to simultaneously smell it all cooking, to be able to look outside and watch Jacques scooping up the grain that made this animal taste and look the way it does. To be able to poke my head out of the cutting room and smell the live pigs growing and gestating and eating and shitting just a few feet away. I could feel my brain growing new pathways as it all unfolded in front of me.

Marc placed his left hand on one of the handles of a sturdy white plastic tub filled with cooked meat, which I assumed came from the heads they'd cooked that morning. He motioned for me to grab the other handle and follow him out of the room. As we made our way back into the main cutting room, it felt as if Marc was pulling me by this boat of meat and I was water-skiing over the fat-slicked floor.

The cramped room we took the meat into—more of an alcove, really—had two machines in it. A grinder and a mixer, I deduced, after Marc did a little miming for me.

Into the mixer we poured the lug of cooked, tender meat, plus a pot of chopped, cooked leeks and onions. I asked Marc what part of the animal the meat came from, and he pointed to his cheek and the hollow just below his eye. Back at Camont the next day, Kate confirmed that they did in fact use meat from the cheeks and around the eye for their blood sausage, plus a little skin and other offal like the lungs, spleen, or stomach. Marc sprinkled in salt and ground black pepper and then stepped away from the mixer. *"Un moment,"* he said and disappeared, reemerging with a blue bucket full of blood, which he placed at my feet. The blood was dark and thick, but still liquid, the result of that woman at the slaughterhouse catching it in a bucket as it spilled out of a hole near the pig's throat and stirring it to keep it from coagulating.

Marc touched my shoulder and then motioned to the bucket of blood with a nod of his head. He said something in French and I noticed, for the first time, that he had a subtle lisp. I stared at him and waited, but he did nothing. He meant for me to do it. This was my job now. So I strained to lift the bucket, I'd guess a good twenty or thirty pounds, and rest it on the lip of the mixer. Marc motioned for me to pour it in.

"Slow," he said in English.

You might expect me to stop here and talk about how horrifying the blood looked as it spilled into the mixer. Perhaps I'd gag a little, or become weak in the knees. But this would serve only to abstract this very simple ingredient—an essential, edible part of an animal raised for food—into something scarier than it actually is. And I did not travel to France with abstraction in mind. I came to France in search of the thing itself, the genuine article. In the Chapolards' *salle de découpe,* the act of cringing felt like the very opposite of understanding. That bucket of blood helped me to understand how cringing can prevent us from thinking or feeling. For what is a cringe but an attempt to separate ourselves from the animal world, from the not-so-pleasant ways some of our food gets to our tables? It also separates us from the world our grandparents or great-grandparents lived in, a world that has, per Freud's classic definition of the uncanny, become alien to us after a relatively short process of repression.

So I did not cringe. Instead I watched, mesmerized, as the red blood, the cooked, brown meat, and the white fat slowly bound together and the mixture turned a deep crimson. The tinny, sharp smell of blood and the sweetness of the meat and fat hit my nose but did not overwhelm it.

Bruno and Dominique retrieved a bucket of very large casings—I would later learn that these came from the caecum, the anus end of the intestinal tract, and are called bungs in English and the rather more charming *culs-de-sac* in French—and set up a small machine on another table. They threaded one long casing onto a hollow tube, then stuffed the canister the tube was attached to with our meat/blood/fat mixture. Bruno pressed his knee into a pedal below the machine that forced the meat down the tube that the intestine had been threaded onto, and it filled, like a tubular meat balloon, creating one long, slightly misshapen link of blood sausage. The image of these sausages replaced the raw image of the blood from just a few minutes earlier, and my mind struggled to catch up and adapt. This was a live animal. Now it's dead. Here's its blood. Watch how it disappears into the meat and fat in the mixer. Now we case it. Now it is sausage. Then we will poach it. Then we will eat it.

Marc disappeared into the refrigerated room where they stored their finished product and emerged with a knife and a cooked *boudin noir*, perhaps left over from last weekend's markets. He cut off a large, inch-thick hunk of it and handed it to me.

As I took it from his hand, I pictured the particular way the blood poured out of the bucket, how thick it appeared. I knew that the blood in the piece I held had been cooked and was no longer liquid, and, besides, I loved blood sausage, but I briefly imagined the thickness of that liquid blood in my mouth, and my stomach turned ever so slightly. Marc looked at me. I was meant to eat this piece of blood sausage. This was the test.

Bruno and Dominique looked up from the task at hand and watched me, waiting to see what I thought of their *boudin noir*.

I was disappointed in myself. I wanted to enjoy the blood sausage, the fruits of our labor. I wanted to fully embrace what it was like to be this close to the seeds *and* the sausage, the blood and the bone all at once. What was this minor form of disgust that had settled over me? I had prided myself for so long on being able to eat just about anything, anywhere, at any time.

I took a big bite of the *boudin noir,* its texture, soft, rich, almost like a thick butter frosting. It tasted like sweet metal, wet earth, and cinnamon, although I hadn't remembered us adding any spices. Complicated and delicious. It was perhaps the best blood sausage I'd ever tasted, and yet I did not want very much of it. That is, perhaps, what taking part in the process does to you.

"*C'est bon?*"

I nodded.

"*C'est vrai,*" I said.

It was true.

TWELVE

B y the time Dominique, Christiane, and I arrived at the *salle de découpe* the next morning, Marc had already loaded a van with large coolers full of product for the market in Casteljaloux. Behind the truck he'd hitched their portable meat counter, folded up, on wheels. We waved goodbye to Marc and then it was on with the boots and butcher's coats and hairnets. Bruno, Cecile, and Marjorie were already hard at work.

Dominique turned to me. "Today we cut," he said, motioning for me to follow him into the walk-in, where about half of the remaining sides still hung by their back legs. The other sides had magically disappeared, presumably off to the market in the form of fresh cuts and cooked charcuterie, or perhaps waiting to be salted or smoked or hung to dry in the form of *saucisson* or hams.

In the walk-in, Dominique crouched down to position his right shoulder underneath the middle of a hanging side of pig. He grabbed the front leg, stood up and pushed the front half of the carcass toward the ceiling with his shoulder, then pointed to where the back leg hung from a large metal hook above my head. I reached up and attempted to push the foot up and off of the hook, but the leg was incredibly stiff, so the task felt a bit like I imagined it might feel to navigate down a curved waterslide with a mannequin in one's arms. After a short struggle, the weight of the back half of the side fell quickly toward the floor, but Dominique shifted in such a way that the midpoint of the pig rested on his shoulder, with one pig foot sticking out in front of him and one behind. He calmly exited the walk-in and moved over to the table Bruno was working at. Bruno set down his knife and helped Dominique slide the side onto the table.

"Ready?" Dominique asked me in English.

"*Oui,*" I said.

"But first," he said, grinning, "stand up straight." Dominique put his hand on my back. I was slouching.

"Breathe," he said, taking an exaggerated breath in and out and then motioning for me to do the same.

"And smile," he said, placing his index fingers at the corners of his upturned mouth.

Dominique was, in so many ways, the antithesis of the kind of butcher I'd had in my head before I came to France. He was soft-spoken, always smiling with his eyes, the epitome of patience and kindness. I could not detect the slightest hint of ego in his voice. No machismo. No boasting. No competition. No judgment.

Before I went to France, I'd sought out a few butcher shops in Portland, what few we had at the time that were not grocery store meat counters. They were mostly lorded over by big men with tattoos and scowls, who answered "No" when I asked them whether they'd ever be willing to teach me. I didn't even know then how very little butchery any of them actually did. Perhaps that was why they brushed me off, but I mostly assumed they just didn't think I looked the part. Smiling and breathing and standing up straight with Dominique, I felt grateful that

I had not wasted too much of my time trying to convince those men to take me under their wing.

"We begin," Dominique said, and then he picked up a short, curved blade with a green handle. His grip seemed vulgar. Wrong. Backward from how I had learned to hold a chef's knife. My only reference for such a grip was from the movie *Psycho*, the scene where we see Norman Bates's silhouette behind the shower curtain holding up the knife and then striking downward at his victim. It was a psycho killer grip. I hate admitting that "psycho killer" was what ran through my mind in that moment, or that, had I been able to express it in French, I might have even made a joke about it, a joke that would have seemed wholly inappropriate and so very American to Dominique. But I'm admitting it here in order to bring to light another form of abstraction many of us are guilty of when it comes to thinking about how meat gets on our tables. By resorting to a horror movie metaphor, I was conjuring a moral judgment, and yet that moral judgment did not stop me from eating meat.

Thankfully, the series of steps that Dominique performed next was so far from "psycho killer" that it was difficult to believe that my mind could ever have gone there. "Grip" didn't even seem like the right word. Instead, Dominique gently guided the tip of the knife into the seams between muscle, fat, skin, and bone as if wielding a divining rod.

Much later, when I began working with career butchers back home, I learned that this fairly universal "butcher's grip," as it is called by those in the industry, allows butchers to keep their wrist straight while working on a carcass, thus preventing awkward positions that could result in tendinitis or carpal tunnel. But Dominique's grip was more than just a form of bodily preservation. It was an eloquent means of reading the map that the pig's anatomy gave him.

DOMINIQUE POINTED TO an area just below the curved back portion of the spine, above the belly and between where the ribs ended and the back leg began.

"*Filet mignon,*" he said. Tenderloin.

The only times I'd ever heard or seen the words *filet mignon* were in relation to beef, and it was always a very expensive item on the menu. Usually, it came on a plate in the form of a "medallion" and would be served with some kind of cooked vegetable, a starch, and a viscous brown sauce drizzled over it. It wasn't something I ate at home as a kid, and we never went to restaurants that had it on the menu. By the time I was in college I thought of it as a cut that rich men in business suits ate. It wasn't until I got into food writing that I began eating filet mignon, although not usually on purpose. It was most often brought to me as a gift from the chef, or incorporated into a special media dinner, proof that the restaurant here had access to expensive cuts. I hadn't really spent much time thinking about what made such a cut more expensive than others, but I could never figure out what all the fuss was about. Filet mignon seemed boring to me, a dense hunk of meat with little flavor or texture.

I watched as Dominique gracefully flicked the very tip of his knife into the space between the muscle and the length of vertebrae that the muscle adhered to.

"I liberate the muscle like this," Dominique said. Not butchering. Not cutting. Liberating.

Kate walked into the cutting room just then, dressed in a white butcher's coat, hairnet, and boots.

"*Bonjour!*" she sang out, filling the room with her big presence.

"Bonjour!" everyone sang back. Knives down, six sets of kisses from each of us on each of her cheeks, six sets of *ça va*s and *ça va bien*s.

"Oh, good. He just started," she said. "How are you doing?"

"I'm so happy you're here," I said. "I have so many questions."

The filet was covered in lumpy fat, glands, and sinew, and once Dominique had liberated it completely, he began trimming all of that away, until the tenderloin became a pale red tube of meat, a few inches in diameter and maybe a foot long.

I asked Kate about filet mignon.

In France, she told me, *filet mignon* most often referred to pork tenderloin, while *filet de boeuf* or *filet mignon de boeuf* referred to beef tenderloin. How Americans had come to use *filet mignon* for beef, she couldn't say.

When the Chapolards didn't sell out of their tenderloins, Kate said, they turned them into *filet sec*, salted and dried tenderloins, which they sold for nearly double the price of fresh tenderloin. Given the popularity of tenderloin back in the States, I found it surprising that they would ever not sell out of fresh tenderloins.

"Is tenderloin considered fancy here like it is in the States?" I asked Kate.

"Yes, especially at restaurants. I've always wondered, though, why I see a lot of grandmothers buying small amounts at the Chapolards' stall," Kate told me. "Maybe because they are fancy grandmothers," she laughed, "but it may be because tenderloin doesn't require a lot of chewing and isn't too strong in flavor."

Wasn't flavor what everyone was in search of? Why would we pay more money for a cut of meat that wasn't very strong in flavor? How was it that tenderloin could simultaneously be the favorite of French grandmothers with fragile teeth and sensitive taste buds and a symbol of rich American businessmen in power suits?

Dominique set the *filet mignon* aside and moved on, this time to the back leg, the *jambon*.

"Now I liberate the 'am," Dominique said, rendering the *h* in *ham* silent. As he said this, he pointed to his own butt and thigh. Then he placed the edge of his knife between two of the pig's vertebrae, near where the spine curved up toward the pig's tail, cut straight down, then ever so slightly to the left, moving his blade through some kind of cartilage, before slicing straight down again, all in one graceful motion. He lifted the entire back leg up off the table, hock and trotter still attached.

"Et voilà! Jambon!" he proclaimed, wide-eyed, with a look of wonder on his face that he had not had when he cut the *filet mignon*.

DOMINIQUE DISAPPEARED INTO the walk-in with his ham. When he returned, he lightly carved a line right underneath the outer edge of the ribs with the very tip of his knife.

"He's about to take off the ribs and spine in one piece. He'll use his knife to peel the bone off of the belly, loin, and shoulder."

Dominique gently ran his knife, over and over again, underneath the ribs that sat on top of the belly, following the same line he'd created before, finding the natural seam that lay there, a topographic anatomical road map. He used his non–knife hand to push the plate of ribs away from him so that he could see where his knife was going. With just a half-dozen strokes, Dominique liberated all of the ribs from the belly to the point where the ribs met vertebral column and loin.

Dominique spoke in French.

"You should think of this bone structure like a key," Kate translated. "He's released the ribs from the belly, but until he releases the ribs that sit on top of the shoulder, and until he releases the chain of vertebrae that the ribs are attached to, you can't turn the key."

Dominique guided his knife underneath the ribs that sat atop the shoulder and then curved along the side of each vertebra, careful not to poke his knife into the loin, where, Kate told me, pork chops and roasts came from.

"Now I open it like a book," Dominique said.

With his non–knife hand, he opened the rib cage away from the pig's body in the same motion one might use to open a book, as if the ribs were the pages and the vertebrae the book's spine.

"It's so beautiful," I said, marveling not only at Dominique's graceful movements but also the splendid curvature of those bones.

Kate explained that the Chapolards would cleave the ribs, which had very little meat on them, since Dominique had left most of the meat on the belly, into manageable pieces, and people would buy them to make stock, soup, or stew. The French weren't into ribs like Americans were.

"With these bones removed, he can easily see where the shoulder and the loin run into each other and retrieve the entire loin muscle. Aside from what they use for paupiettes and *boudin blanc,* he'll keep the loin muscle whole for the market, and people can ask him to cut whatever size loin roast they want. On the other side of the pig, they'll leave the ribs and spine on and turn the loin into bone-in pork chops."

Dominique pointed to the top of the shoulder, where the ribs and vertebrae used to be.

This cut was commonly called the coppa in Italy, Kate explained, a term the French had adopted, although Dominique sometimes also referred to that area of the shoulder as the *échine.* Dominique motioned for me to examine the structure of the coppa. It wasn't one muscle, but rather a bundle of small, elongated muscles connected by a star of intermuscular fat running between them.

"This is a very flavorful cut, because all those muscles worked very hard and all that fat provides flavor," Kate said, explaining that the Chapolards salted and dried the coppa whole and sold thin slices of it for a high price. The rest of the shoulder they'd use for sausage, brochettes, and pâté. To remove the coppa, Dominique carved a straight line with his knife, starting at the exact spot where the darker-colored coppa met the lighter-colored loin and ending down toward the hock and trotter. Then he used his knife to peel the loin muscle that ran along the pig's back away from a thick layer of fatback and skin and held the lean, tubular muscle up for me to see. It was about as long as my legs and almost as thick as the diameter of my head. Dominique pointed to the area on either side of his spine and said, *"Longe."* Loin.

As Dominique pointed to the muscles of his body, I became more aware of my own. Butt, thighs, calves. I use them for walking, for running, for sitting, for squatting. I use them nearly all the time, as does a pig, at least one that is allowed to move around. What about the loin— the backstrap, as hunters often call it—the long muscles that run along each side of my spine and hold me upright—do they hold a four-legged animal up in the same way? Not quite. Did *how* these muscles on a pig move have something to do with the pale color of the loin and the darker hue of the coppa, with why we cook a pork chop in a frying pan and a pork shoulder in a smoker or a Crock-Pot? I realized that this was the first time in my thirty-two years of eating (and not eating) meat that I had ever given much thought to the intricacies of a pig's anatomy in relation to my own.

"So that is the first step in butchery," Kate said.

"That's only the first step?"

"In the United States, this is what they call cutting primals," Kate said. The shoulder, the back leg, the belly, and the loin were considered the four main primals. There was also *le cinq*, the fifth quarter, as Kate called it: the head, the hocks, the trotters, the tail, and all of the offal, much of which went into the Chapolards' *pâté de tête* and *boudin noir*.

THIS WAS ALL PRIMAL, I thought. A man, a knife, a pig.

Michael Pollan, in *Cooked*, writes, "In ancient Greece, the word for 'cook,' 'butcher,' and 'priest' was the same—*mageiros*—and the word shared an etymological root with 'magic.'" Pollan also posits that a *mageiro*, in ancient Greece, referred to a man—it could only be a man back then—who was hired to kill animals for sacrifice and then roast them for sacrificial public gatherings. In this sense, he was a priest, a gatekeeper between animal, human, and gods, and he was a cook, charged with feeding the public with fire-roasted meats. His tasks also extended to selling whatever meat was left after the feast, so he became a butcher, too. It seemed to me that the *mageiro* had to possess a great deal of respect and reverence to ensure that one of his roles did not overtake the others. The way we produce, handle, and consume meat today is very much the opposite of that. Respect and reverence are almost entirely missing from the equation.

Watching Dominique made me feel a part of something bigger, something full of respect and, yes, maybe even sacred. Underneath this feeling, the stories I had told myself about my relationship to the animals I ate began to smell bitter and more than a little burned.

THIRTEEN

Back at Kate's, there were beds to be made, dishes to be washed. Then her students called to say their flights were canceled and they'd be a few days late. This gave Kate, Jonathan, and me an opportunity to figure out how we would hang shelves on the extremely uneven walls of the piggery the next day. After we hatched some half-baked plans, Kate and Jonathan made us what Kate called a *grand aioli* for dinner, a gorgeous, Provençal-inspired spread of poached carrots, beans, and potatoes along with perfectly soft-boiled eggs from Kate's chickens, their yolks a vibrant burnt-yellow; bright-orange langoustines boiled to tender perfection; and a homemade aioli spiked with garlic and garden herbs for dipping it all into.

When the post-dinner Gascon lull began to settle into my head and feet, I found my hammock for the evening and brought with me a small glass of Armagnac and a copy of John Berger's *Pig Earth*, his 1979 paean to the last surviving vestiges of French peasant culture in the twentieth century, a book he wrote after abandoning his city life for the French countryside, just as I had decided, at least temporarily, to do.

In the book's introduction, Berger describes peasants as "those who work on the land to produce food to feed themselves" but who are "forced to feed others first, often at the price of going hungry themselves. They see the grain in the fields which they have worked and harvested—on their own land or on the landowner's—being taken away to feed others, or to be sold for the profit of others."

The nature of the Chapolards' work had a strong whiff of peasantry—the work of producing food was steady and consistent, based, to some extent, on the seasons—but it seemed to me that the Chapolards had also managed a modern-day workaround. They were selling the food they raised to others, sure, but no one else but the Chapolards appeared to profit directly from those sales—save for the tax bureau, a certain kind of politician might point out. They were intimately involved in every part of the process of getting their product to market, but that product also filled their own tables. The Chapolards were not going hungry—far from it. To achieve this total ownership, they had to work together as a family, and they had to work hard, but none of this came at the cost of their own well-being, at least not so far as I could tell. It was as if they had managed to preserve the fulfilling parts of peasant life—constantly working to feed themselves and, one might argue, their community—and to bypass the parts that no one should rightly be nostalgic about: exploitation by others or, worse, exploitation of one another.

Working in the cutting room was cold, serious, efficient work, but I hadn't yet detected the sort of dead-eyed monotony one might assume would result from this sort of repetition over time. On the contrary, each day I spent in the cutting room, I witnessed a sustained alertness in the eyes of each of the Chapolards, even if their eyes also expressed tiredness. It seemed to me that this sort of sustained energy could only be the result of pride in one's work, and that that pride might have something to do with their particular mode of ownership of the production process.

"Recently the insulation of the citizen has become so total that it has become suffocating," John Berger wrote. "He lives alone in a serviced limbo—hence his newly awakened, but necessarily naïve, interest in the countryside."

Berger wrote this in 1979, but, reading it, I felt like he was talking about me. Born in 1976, perhaps I was the progeny of that 1979 nostalgia. I most likely *was* romanticizing the Chapolards' way of life and work, but perhaps my romantic projection came out of a longing for purpose and meaning amid my own serviced life back home. I had a hunch that *Pig Earth* represented Berger's own attempt to reconcile a similar nostalgic longing.

A FEW WEEKS LATER, over lunch one day, I decided to don my old magazine editor's hat for an hour, convinced Kate and a few of her students who spoke French to translate for me, and began peppering Dominique and Christiane with questions. Did they think someone like me who came to study with them was just consumed by useless romanticism?

Christiane responded first by reiterating that they worked hard every day, and that she liked best those visitors to the cutting room who were willing to pick up a knife and work hard like them. But she also said that she thought maybe some people didn't realize how much their family had struggled in the beginning to figure out *how* to work together. They'd even hired a psychologist at one point to help the brothers and their wives figure out how to cooperate, negotiate, and still get along at the end of the day.

When talking about the importance of working together, Dominique often liked to say, *"Tout seul, tu meurs"*—in other words, alone, you die. But that didn't make working together any easier, Christiane said.

Then Kate explained something I hadn't understood at first. Not only had the Chapolard brothers and their wives formed a Groupement Agricole d'Exploitation en Commun (GAEC)—a kind of legally recognized business status that was first created and recognized in France in 1962 to support and sustain the family farm model and prevent rural exodus—by obligating family members to formally and legally share in the liabilities, risks, and rewards of running their operation together. They also took part in a Coopérative d'Utilisation de Matériel Agricole (CUMA), which translates loosely as "a cooperative for the use of farm equipment." Taking part in a CUMA meant that the Chapolards shared the cost and use of farm equipment with other local farmers so that no one farmer had to share the financial burden of such investments. It was this same principle of shared resources and risk that allowed the Chapolards to team up with other farmers and cooperatively take over the abattoir in Condom.

Dominique called their particular method of cooperation and vertically integrated production "short-circuit farming." Kate, on the other hand, often called it "full-circle farming." Growing their own feed controlled grain prices, and transforming the meat themselves into value-added products like sausage and *jambon* meant the profits stayed with the family. Forming a GAEC and joining a CUMA meant the individual costs and risks of starting and running such a farm were kept to a minimum, and it allowed them to remain part of a community of farmers they could learn from.

This wasn't a choice that most farmers in France made, but Dominique explained that it had been part of what appealed to him when he decided, at the age of forty-four, to quit his job as an administrator at a forestry school and start working with his brothers on the farm. He wanted to prove that small-scale family farmers could succeed, but only if farmers supported one another, and only if they lived in a community that was willing to support them with their pocketbooks. The meat the Chapolards sold never traveled farther than fifteen miles from their farm, and they sold every part of their animals—to a population of fewer than ten thousand people.

He then explained just how much work and time goes into their process. It takes six to eight months from the time they sow the grain seeds in the spring to when they harvest the grain and feed it to their pigs in the fall. From breeding to farrowing to weaning to fattening to slaughter, another sixteen months passes. They spend every week in the cutting room working on eight to ten pigs at a time, each weighing an average of almost four hundred pounds at slaughter. Their charcuterie takes four to eight weeks of curing and drying. They spend four days a week at the markets. Add all of this up and their seed-to-sausage process represents over two years of investment before they receive the nine euros per kilo, on average—roughly five dollars per pound—that they charge today for their products.

Christiane chimed in again. "This isn't a hippie thing," she said. "This isn't a back-to-the-land, peace-and-love ideal. We work hard and we own what we do. But nothing is permanent. We allow ourselves moments of pleasure and joy because our life is day by day and we don't know what is going to happen tomorrow."

"To a peasant," Berger wrote, "the future is . . . [a] narrow path across an indeterminate expanse of known and unknown risks. When peasants cooperate to fight an outside force and the impulse to do this is always defensive, they adopt a guerrilla strategy—which is precisely a network of narrow paths across an indeterminate hostile environment."

I believe this is exactly what Christiane meant by not knowing what was going to happen tomorrow, and what Dominique meant by his you-work-alone-and-you-die motto. The Chapolard family had employed its own kind of guerrilla strategy, a shared network of precarious paths across an ever-changing, often hostile economic and cultural environment. To some, this might seem like a burdensome amount of work. Why use a boning knife when a band saw is so much quicker? Why raise an animal humanely and slowly when it takes so much longer, and no one will even notice the difference anyway? But to me, the Chapolards' guerrilla strategy and all the work it required of them seemed defiant. And maybe that kind of defiance was what I'd been missing in my life.

FOURTEEN

Back in the *salle de découpe* on Friday, Christiane instructed me to join Marjorie in stringing cubes of meat onto skewers for the next day's market in Nérac. I set myself up so that I had a view of Bruno's worktable, where he gracefully ran the blade of his knife along the contours of a few bones on a primal he was working on, removing what looked like a leg bone and maybe a shoulder blade. Since one of the muscles on the table looked a bit like the coppa I'd seen Dominique extract from underneath the ribs a few days back, I concluded that he was deboning a shoulder primal.

Every few minutes, Bruno would swipe each side of his knife very lightly along a honing steel. I'd watched so many chefs do this back home, but it was always with bravado—they'd hold the steel in front of their chests when they did it, drawing the knife dangerously toward them, and it was always very loud. By contrast, Bruno set the tip of his honing steel on the table to steady it, and his swipes were so light that I strained to hear whether the blade was actually touching the steel.

Bruno disappeared into the walk-in, then carried over to me a messy, tight mass of intertwined muscles that looked a bit like the primal he'd been working on. He picked up a few of the pieces of skewer meat I'd been threading onto skewers, then handed me a knife, saying something incomprehensible in French and pointing to the morass of meat he'd just set in front of me. I was unsure of what he wanted me to do and had no idea how to ask him to clarify, so I assumed he wanted me to cube the meat up for more brochettes and that it didn't matter much how well I did it. I committed to a place for my knife to go and proceeded to turn the muscles into cubes.

After a while I looked at Marjorie. *"Comme ça?"* I asked her. Like this? She nodded silently—I had yet to hear her utter a word.

Next to me, Dominique massaged handfuls of coarse, grayish salt into several tightly grained muscles. Christiane tended to her pâté in the back room. Cecile tied roasts with red-and-white butcher's twine.

Eventually, I built up a small mountain of brochette meat on the table and was just about to show Bruno when Marc walked in, looked down at my work, and yelled, *"Merde!"*

All six people in the cutting room began speaking in quick French.

I looked from one mouth to another as a young child might. I sensed the tension, knew it was aimed at me, but had no idea what I'd done wrong, so I continued cutting. Bruno motioned for everyone to be quiet, then moved toward me in slow motion. He put his left hand up gently, to stop what I was doing, placed his other hand over the handle of my knife, and took it away from me as a cop might cautiously take away the gun of a bank robber once she'd surrendered. And then he pulled me by the arm into the cold-storage walk-in, where a few primals and one side were still hanging. He motioned for me to crouch down near the floor with him, grabbed my right hand, and moved it toward the front leg of the hanging side. Then he moved my right hand to my own shoulder. *"Épaule,"* he said, motioning for me to repeat him.

"Épaule," I repeated.

We stood up together and he moved my right hand toward the pig's back leg.

"*Jambon*," he said.

"*Jambon*," I repeated.

"Expensive," he said.

"Expensive? Shit!"

"*Merde!*" Bruno exclaimed, offering me a tense smile.

I realized, finally, what I'd done. I'd mistaken the ham for the bottom part of the shoulder, and transformed one of their most expensive cuts at the market into cheap skewer meat for the grill. Bruno had clearly assumed I knew what I was doing, when I barely knew a pig's ass from its head. Had I not turned this meat into cheap brochettes, they would have rubbed each of the little hams—which they called *noix de jambon*—with that coarse gray sea salt I'd seen Dominique using, netted them, and then rolled them in copious amounts of cracked black pepper before cold-smoking them overnight in their "smokehouse," which was really just an old fireplace in a crumbled part of their parents' *maison*. Afterward they would have brought them into the drying room for nearly a month, until the hams shrank and became the most delicious little smoky *jambons* I'd ever tasted. They sold these at market for several times as much as they sold their brochettes, and I'd screwed it all up.

Why had he handed me this ham as though I knew what to do with it? Was it a test, like the blood sausage? Was it Bruno's mistake or mine? Why had Marjorie nodded in approval? Maybe I had a way of convincing them that I knew exactly what I was doing when in fact I could be sure of very little.

"This pig, I open it like a book," Dominique had said. This book was in two languages I didn't understand—pig and French—with no glossary at the back.

Bruno put his hand on my shoulder and nodded silently, then motioned for me to follow him back into the only slightly warmer cutting room.

"*Je suis très désolé,*" I said to everyone, my face red with shame. I'm so sorry. It was one of the few things I could confidently say in French at this point, besides "I don't know" and "What is this?" But they had already moved on. There was still work to do. Dominique came up next to me with a finished *jambon* in his hand and a knife.

"Camees," he said—the Chapolards always seemed to pronounce my name as if they were attempting to sound like an American who was attempting to sound French—cutting into the ham and offering me a slice, "dees is whaht we do wit *jambon*." He capped it off with a smile that his mustache only served to accentuate.

I felt guilty eating the piece of ham Dominique offered me. The Chapolards had taken me under their wing for very little money—I'd merely paid for them to feed me lunch and dinner a few times a week—and I had probably lost them all of that money and more. But I took a bite anyway, knowing it was expected of me, that this was part of the lesson, and when I did, my face flashed hot again and I suddenly felt like weeping. I'd come to France feeling like such a failure, and this single mistake—which, considering I'd been cutting pork for only a week, was fairly forgivable—felt, in the moment, like further evidence of my inadequacy.

I'd tasted the Chapolards' ham before, but today it tasted entirely new. The *jambon* was salty and smoky, as always, but it had another flavor now. As a young and inexperienced food writer, I would have used the word *umami* to describe it. But that word seemed like a black hole into which my more complex experience of flavor and emotion could disappear. It wasn't just flavor. It was something else altogether, and it needed a new word. But I could think of nothing.

Dominique watched me chew and swallow, then he said something in French that I'd heard him say a few times before. It sounded something like *"Sayy-for-mee-dawb-luh."*

Before I could ask Dominique what he meant, Christiane touched my shoulder and said, "We go now?" Which meant it was time for her to drive me back to their house so that Kate could have lunch with us and take me home. It was to be a short day for me.

DIRTY RAGS AND SMOCKS filled the passenger seat of Christiane's beat-up Peugeot, so I parked myself in the back. We were silent as we drove away from the farm, down the long, narrow dirt-and-gravel driveway, bordered on either side by the golden barley and triticale wheat they grew to feed their pigs. I was suddenly tired. Once again, I felt like a burden, a disappointment, a failure. A fairly understandable mistake had brought the pendulum of loss and doubt from the past year crashing back toward me once again. What was I even doing here? How could I be so stupid? Shouldn't I be back home finding a real job, settling into life with Will?

We bumped along the dirt road toward the highway, kicking up dust behind us, and then Christiane came to a sudden stop, turned the car off, and whipped around in her seat.

"Camees," she said. She took a quick breath in and paused, perhaps trying to figure out if she should speak in English or French, and then began speaking in French, albeit slowly, for my sake. Given the fact that I clearly had no clue what anyone was saying in French, whatever I heard her ask me could have been completely wrong. Maybe she asked me if I needed to go to the bathroom. But this is what I think she said, because this is what I needed her to say in the moment:

"What will you do with all of this when you go home?"

There was so much I could have said: I'll make bacon! I'll work in a butcher shop! I'll start a farm like yours!

Instead I said: *"Je ne sais pas."* I didn't know how to say what I wanted to say. But I also didn't know the answer to her question. Maybe I wouldn't do anything with all of this. The last time I had done something of significance with my life, it had collapsed in on itself.

Then she said, "If you don't do anything with it, we will be very sad. It will all be for nothing."

BACK AT Kate's *pigeonnier,* I told her what had happened over glasses of sweet Floc. The sun was setting and, as it went every evening in the waning light of that blue hour, Bacon paced back and forth across the herb garden outside Kate's kitchen window, barking.

"What does *Sayy-for-mee-dawb-luh* mean?" I asked Kate.

"What?" she asked, laughing at my ugly accent.

I repeated it for her.

"You mean *'C'est formidable'*?"

"Yeah, I think so," I said, slurring a little from my third glass of Floc.

"'That's formidable,'" she said.

"What is?" I asked, confused.

"That's the direct translation," she said. "'That's formidable.'"

Not unlike the verb *cleave,* which can mean to divide or split but also to adhere closely or loyally to, *formidable,* an adjective, has contradictory definitions. If something is formidable, it can inspire fear, dread, or apprehension. But it can also inspire wonder.

Dominique had been describing the flavor of that ham, of course. But more than flavor, the word perfectly described my uselessness, the uncertainty I had about everything in that moment, as well as the uncertainty I would face when I returned home.

Christiane's words lodged into the tiny space between my rib cage and my heart.

FIFTEEN

At the end of my second week working with the Chapolards, Kate drove me to the Nérac market, where I would spend the day working alongside the Chapolards at their booth. From Camont, we drove west on the autoroute and then turned left at the sleepy town of Bruch, where the owner of what appeared to be the town's only restaurant was setting up for the day, dragging bistro tables and chairs out front. We kept going, past more fields of rapeseed and sunflowers, until we passed through another sleepy town, Espiens, and then turned into a busy two-lane roundabout that shot us out toward the Petite Baïse river, which splits the town of Nérac in half. From the river I looked up at Nérac's many stone buildings, perched on the hill above the water. The same burnt-cream color graced most of the buildings, but their owners had painted their front doors and shutters forest green, indigo blue, and an occasional cherry red.

Kate navigated the car around the perimeter of the sprawling outdoor market and parked, and we headed to the Chapolards' stall. I stopped to take a picture of every colorful door we passed, many with their own unique brass or steel knockers. A hand. A tiger. A golden sphere. A horseshoe. Lucky charms and totems. I wanted to knock on each door and meet the people within, to enter their worlds and leave my own world back home behind me.

"*Bonjour!*" Kate yelled to Dominique, who was sporting his market beret and looking, in every way, quintessentially French. Dominique, Christiane, and their daughter Mathilde, home for the summer from university, were busy putting the final touches on their portable meat case, which they'd set up underneath a white tent. Behind them, a bright-yellow banner read DU PRODUCTEUR AU CONSOMMATEUR L ART DE COCHON. "From Producer to Consumer—the Art of the Pig." Their meat case rivaled that of any indoor butcher shop I'd seen in France so far.

"It took me years to get the double entendre," Kate said. "*L'art* and *lard* sound the same when you say them."

There were several other outdoor meat counters at the Nérac market, some much bigger than the Chapolards' setup. Most were producers like the Chapolards, specializing in one species, while a few were more traditional, albeit portable, butcher shops, buying animals from farmers like the Chapolards and then butchering and further processing them for their customers. Some specialized in only cooked and cured products, like Jehanne and her foie-gras-stuffed duck prosciutto.

At the farmers' markets in Portland, as well as those in most other American cities I'd been to, plenty of meat producers set up booths to sell their products, but food safety regulations can make it a challenge to sell fresh meat at the markets. And since farmers often are not inclined to re-portion the meat they are selling once it has been cut and wrapped at the slaughterhouse or other processor, they mostly sell frozen cuts that have been vacuum-sealed or paper-wrapped, placing them in coolers full of ice and putting chalkboard signs out to advertise what they have for sale. This makes buying meat at the farmers' markets in

America a crapshoot. If you ask the farmer to tell you about the meat, most of them will talk your ear off, but there is very little sensory activity involved in buying meat from them. Plus, the farmer is almost never the person who killed the animal or butchered the meat. So you glean what information you can, you buy the meat, you hope it's good, you take it home and cook it, and then you decide whether you're going to go back and buy more.

Before the market officially opened, Dominique and Christiane walked me through the case. One side was taken up mostly by fresh cuts like the long, trimmed whole loin that they would cut roasts off of when customers asked, plus a few of the one- and two-pound shoulder roasts—at least I thought they were from the shoulder—that Cecile had trussed on Friday. A fanned pile of thick, red, bone-in pork chops, each with a healthy layer of fat around the edges, surrounded the roasts. Christiane's beautiful paupiettes lined the front of the case, each tied with twine like a present. Christiane told me that if customers asked how to cook them, I should tell them to brown the paupiettes over high heat and then stew them over low heat with tomatoes and onions. I tried to memorize how to say this in French, in case anyone asked me, but immediately forgot.

With the exception of the paupiettes, these were all cuts that I would likely see at any meat counter in the States. But there were also trotters here. Raw pig ears. Cooked tongues. A few fresh hocks. There were the rolls of skin that Marjorie had neatly tied. No brains today, since the Chapolards had sold out of those earlier in the week. All of these parts were reminders that this meat came from a live pig, a pig with feet, a head, skin, and bone. Unless I went to an Asian or Latino grocery store in Portland, I wouldn't see these parts. In fact, most meat-counter workers tried very hard to make sure that we did not have to be reminded of such things—not only by omitting the parts that look too much like they came from an animal, but also by sprinkling the case with parsley or rosemary sprigs, lemon or orange slices, or those bright-green plastic separators made to look like fake grass.

On the other side of the case, the Chapolards had piled all manner of sausages, including their *boudin noir,* which they'd poached off already, a process that had turned the blood sausage an unattractive brown color, and yet by the end of the day they would sell out. The Chapolards sold the rest of their sausages—a bright-orange merguez, as well as some simple pink-and-white sausages made from just salt, pepper, pork, and fat—uncooked, leaving all the links attached to one another, such that each long snake of sausage links piled into one big haphazard mountain, a good reminder that these sausages were cased in intestines that formed their own unique snakelike coils inside the pig they'd come from. Customers would ask for a certain number of sausages, Dominique explained, and then we would count that number of links and cut them off from the rest.

A few brined, boiled, and pressed hams peeked out from behind the mountains of sausages. I saw two oblong loaf pans of the *pâté de tête* they'd made earlier in the week. In the very front of the case, a few more pans of Christiane's pâté, which sat next to yet more pans of smaller, fist-size balls of that same pâté, each covered in caul fat; they called these *fricandeaux,* and they were good for one or two people, she said. In the case, the Chapolards had also placed a pan of *gratton,* a coarse, slightly fatty meat spread they make by slowly cooking stray pieces of skin, fat, and other scraps left over from the butchery process, until all the fat has rendered and the meat and skin have turned tender. They then whip this all together to form a smooth meat paste, meant to be spread on bread like butter.

Above the refrigerated meat case, Dominique and Christiane elegantly displayed all their cured, dried charcuterie—from *ventrèche* to *saucisse sèche, filet sec,* and the small, boneless, salted, smoked, and dried hams they called *noix de jambon.*

Charcuterie means, literally, a butcher's shop that specializes in pork. It comes from the Middle French word *chaircutier,* meaning "pork butcher" (from *chair cuite,* or "cooked meat"). But the word *charcuterie* has come to refer to all manner of meats, although most often pork, that are preserved through various processes of salting, smoking, drying, and cooking. If you've eaten a bologna sandwich, you've eaten charcuterie. If you've eaten deli ham or hot dogs or bratwurst or smoked fish or salami, you've eaten charcuterie. Charcuterie, when it's called that— as opposed to, say, Oscar Mayer cold cuts—can be quite expensive in the States, and has thus ascended to the realm of fancy food. But it does not have expensive or fancy roots. The charcuterie we eat today, whether it's confit or prosciutto, was born of peasant ingenuity—a means to make meat last before the invention of refrigeration. This ancient, simple formula of salt, meat, air, and time makes up the bread and butter of the Chapolards' operation.

DOMINIQUE AND CHRISTIANE showed me where I'd get change from. They gave me a quick lesson on the slicer and another on how to wrap meat in butcher paper. And then it was time. A line formed almost immediately. Hunched Gascon grandmothers, young Parisian parents on holiday with their tight jeans and wispy bangs, and British and American expats all stood patiently in line. We haggled over trotters and hocks, shoulder roasts and blood sausage. I faltered every time I had to count out change in euros. I'd naturally begin speaking in

Spanish when I didn't know what to say in French. At times I intuitively spoke in English with a bad French accent. To their customers, I became grand entertainment.

The expats gazed into the meat case in awe and wonder and asked us, "What's good today?" to which I replied, "It's all good!" Each part of the pig deserved a place at the table. The Chapolards' customers seemed flexible, curious, mostly confident about what they were buying and why, and the majority of them, save for a few younger people, seemed to know what to do with just about every cut in the kitchen. I also noticed that no single customer bought that much meat. The Chapolards were relying on a lot of customers to buy a little. A bit of salted, dried *jambon* to snack on for the week. Maybe a couple of sausages. A slice or two of *ventrèche* to put in soups or salads. Occasionally a whole roast, but only if they were feeding a large group. There was no stocking up for the week with Costco-style quantities. No one planned on freezing anything. All the meat the customers bought would be used within a few days (save for the dry-cured meats, which would last much longer), and it would be used sparingly.

One older woman approached the pile of *saucisse sèche* and proceeded to squeeze every single one between her thumb and forefinger. She settled on the one she wanted, looked at me, and pointed to it. It was at the bottom, but I managed to retrieve it without toppling the entire pile. It was very soft; in fact, so soft that I wondered if it was safe to eat. Had it lost enough moisture yet? Had it fermented properly? But over the course of the coming weeks, Kate and the Chapolards helped me to understand that it was perfectly safe once it lost around 30 percent of its weight in moisture. Plus, the Chapolards could vouch for this meat. They'd shepherded the *saucisse sèche* through every step of the process. Kate told me she would never buy a *saucisson* that was that soft in the

States, because she'd likely have no idea where the pork came from, how it had been handled, or how old it was, because it would probably be made from younger, more watery pork—because she just wouldn't know, and nobody working behind the counter would be expected to. Then again, she said, you'd probably never find a *saucisson* that soft in the States in the first place, precisely because of that lack of knowledge and the risks that lack posed.

The old woman pointed to the boiled ham and asked for a *tranche*, or a slice, specifying with her thumb and forefinger how thick she wanted it. "Enough for the week," she said. The slice I handed to her was about the size of a piece of ham that might be served on a breakfast platter at Denny's for one person. For this woman, it would last her the week.

Then she pointed to a small pile of trotters, each of which the Chapolards had cut in half lengthwise with a cleaver so that customers could more easily render the fat and gelatin within when they cooked it.

One very soft *saucisson*. One thick slice of boiled ham. One trotter. That was it, her stash of pork for the week. Maybe she'd go somewhere else to buy some other kind of meat. But I assumed it wouldn't be much more than what she'd bought here.

"What's that?" one woman in her midforties, dressed in a flowing white sundress, asked me in a British accent, pointing to the blood sausage.

"That's *boudin noir*—blood sausage," I told her. "I made it myself!"

"Oh, yes. I grew up eating blood sausage. I haven't had it in years. Why don't you give me a few inches."

Every few minutes, Dominique and Christiane recognized one of the customers and stepped down from our raised platform behind the case to say hello and kiss them on their cheeks. "Our customers are our family," Christiane said to me.

Toward the end of the market day, a little after lunch, we ran out of most everything except pork chops. Pork chops! In the States it would most certainly be pig ears and trotters we'd have trouble selling. A few more customers came by, one looking for brains and one looking for a little *gratton*. We had neither, so they each bought pork chops instead. They'd work with it. They'd change their plans. They'd go home to their kitchens and be resourceful.

SIXTEEN

One Monday morning, Bruno and Dominique asked me to help them process pig heads for the first time, a task that required me to wield a cleaver, a tool that had, of late, filled me with dread. I'd watched in the weeks before as Dominique and Bruno effortlessly cleaved straight lines through racks of ribs. When they'd encouraged me to try this out myself, I stood up straight, put on my most confident face, and managed to make several minor dents in various portions of the rib bones, with not even a hint of a continuous line in sight. I also flinched every time my cleaver hit bone, not so much because I was afraid of the cleaver—sure, it was dangerous and could easily cut my finger off with one wrong swipe—but because I was, to put it bluntly, afraid of fucking up.

And yet Dominique, perhaps sensing my hesitation, continued to hand me rib bones. After a few tries, my dents finally turned into cuts and my lines turned from haphazard hacks to something resembling the pattern left in the snow by a downhill skier, but I would probably need another fifty sets of ribs before I got it right. No wonder people used band saws—any freshman could use a band saw and achieve a straight line the first time around.

Bruno and Dominique applied cleaver to skull in a slightly different manner than they did to ribs. They'd each place a head at the corner of the table, right above where the table leg met the tabletop—the sturdiest part of the table, they explained—then they'd draw the cleaver up into the air and strike down right between the two ears. The cleaver immediately lodged where it was supposed to, but because it was too difficult even for them to pull the cleaver back out immediately and go for a second swing, they'd adjust their grip on the cleaver's handle with their knife hand, grab on to the skull with their other hand, pull the skull and the lodged cleaver up into the air together, and bang it down on the table. They'd do this a few times, until they'd cleaved about halfway through, then they'd pry the cleaver out, pull the skull apart as best they could with their hands, and, to finish the job, gently and rhythmically hack away at the crack's center with the cleaver, using smaller chopping motions, until the head split in half completely.

No one wore protective gloves in the Chapolards' cutting room—Dominique once said something to me about how wearing protection makes you more careless, which seemed blurry in its logic. The day they encouraged me to cleave open my first pig head seemed like the right time to change that policy.

Dominique started the pig head for me, lodging the cleaver into the middle of the skull. Then I grabbed the cleaver handle with my right hand and the skull with my left and heaved the entire thing up into the air. It was heavier than I thought it would be. The cleaver alone probably weighed ten pounds, the head another thirty. I banged the whole thing down loudly on the table, but the cleaver didn't move at all. I tried again, this time lifting it up higher and using more intention and force on the way down. The cleaver moved a little, but it had decided to angle to the left.

Dominique held up his hand to stop me.

"Camees, this is not lovely," he said in English. "I show you."

He pulled the cleaver out of the head and, with a smaller arc this time, lodged it back in, angling it slightly to the right to correct for my mistake. These movements felt so brute and vulgar to me, my tool an ancient, imprecise caveman invention at best, and yet Dominique managed to be so meticulous and graceful in those same movements.

I mustered even more strength and force this time, pulled the whole thing up and then down toward the table, and as I did so I imagined that I was actually pushing the skull through the table, toward the floor. This Jedi mind trick served to change the impact in such a way that the cleaver nearly made it all the way through. My dad had taught me to chop wood this way—"Imagine the ax going all the way through the wood," he'd said, "not just hitting the wood's surface."

"*Voilà!*" Dominique said. "That . . . is lovely." *Lovely/not lovely* had become Dominique's main way of assessing my progress in his cutting room.

I went to pull the cleaver out of the skull, but it was stuck. *The Sword in the Stone* flashed in my mind. Dominique tried to pull my cleaver out, but even he struggled to remove it.

"*C'est très formidable,*" he said, laughing.

Once he managed to pull the cleaver out of the skull, he handed it back to me to finish the job. I gently chopped away at the crevice I'd created with the cleaver's blade, but tentatively—my other hand still hanging on to the skull, right in the path of the cleaver. My arms felt wimpy. But this feeling was more a product of fear than a reflection of my actual strength. I didn't want to hurt myself. I didn't want to screw up the job, either. I needed to commit. I also realized that I needed to do this every day if I was going to get good at it. This was dabbling, I thought. Dabbling would get me nowhere.

It took me about ten minutes to finish cleaving that one skull while Bruno and Dominique finished the rest. By the end, I was out of breath and red in the face.

Bruno and Dominique then showed me how to gently scoop out each of the halves of brain from the skull. The contrast was stark: big men, with thick hands and fingers, cleaving away at a skull, then big men with thick hands and fingers delicately scooping out these tiny brains as if they were newborn birds.

I inspected the two halves of brain from the one skull I'd managed to cleave open. They were covered in bone chips from my clumsy cleaving, but for the most part, the two halves had remained intact. I gently worked my fingers in between one of the halves of brain and the curved cup of skull that it sat in. The brain felt soft and cold. It was the color of coffee with lots of cream in it. I could see the coils and vessels that made up the brain's signature structure. I'd always imagined the brain of any animal to be somewhat durable and hard. But this brain was quite soft and malleable, almost like flan.

As I pulled each half completely out of the skull using two of my fingers, I imagined my own brain in its own curved recesses. It was difficult not to, just as it was difficult to remove the shoulder blade and not feel my own shoulders, sore and tired from working in the cutting room.

EVOLUTIONARY ANTHROPOLOGISTS largely agree that eating meat (along with a side of tubers and honey) made the expansion of early human brains physiologically possible. They believe this because meat, as opposed to more fibrous sources of food, like leaves and fruit, provides more concentrated calories and takes less energy to digest, and so a diet with some amount of meat in it would have allowed our brains to grow and our guts to shrink.

Some evolutionary anthropologists have also remarked on the social implications of meat eating, suggesting that the social aspect of finding and sharing that meat—the communal hunting and scavenging that made eating meat in higher quantities possible—required us to grow bigger brains. Competing and cooperating, creating alliances, and teaching one another how to hunt takes a certain kind of intelligence, after all. Adding meat to our diets likely gave us the intelligence we needed to work together to find more meat to eat, which in turn made our brains keep growing.

So there I was inside the black hole again, standing in the Chapolards' cutting room, holding the two halves of a pig's brain in my palm, feeling my own brain floating in my skull—something I can't actually feel, but nevertheless there I was feeling it somehow—thinking with my big human brain about meat and brains and evolution, about predator and prey, about the fact that, because we figured out how to eat meat long ago, our brains grew big enough to make us capable of questioning the ethics of eating the very ingredient that allowed us to ask such questions.

Pig eyes. My eyes. Pig brain. My brain. Pig tongue. My tongue. Pig skull. My skull. For some people, pressing like against like, as I was doing, standing there with a dead pig's brain in my hand, inspires revulsion. But instead I felt kinship. Reverence. Wonderment. Trepidation. And melancholy. At once.

In his essay "Why Look at Animals?" John Berger has described the uncanny ways in which animals are "like and unlike" humans. In his view, animals resemble man in three main ways. They are born, are sentient, and are mortal. But in their "superficial anatomy," in their habits, their relationship to time, their physical capacities, they are quite different. Because of this, when an animal regards a human, it "scrutinises him across a narrow abyss of non-comprehension. . . . The man too is looking across a similar, but not identical, abyss of non-comprehension. And this is so wherever he looks. He is always looking across ignorance and fear. And so, when he is *being seen* by the animal, he is being seen as his surroundings are seen by him. His recognition of this is what makes the look of the animal familiar. And yet the animal is distinct, and can never be confused with man. Thus, a power is ascribed to the animal, comparable with human power but never coinciding with it. The animal has secrets which, unlike the secrets of caves, mountains, seas, are specifically addressed to man."

These "secrets were about animals as an *intercession* between man and his origin." For us, animals bridge the gap between nature and culture—they represent our past, but also something greater. And thus, animals became our first chosen metaphor for ourselves, in the form of myth and legend, crude paintings on cave walls. These universal "animal-signs" allowed us to chart our own experience of the world.

"Animals came from over the horizon," Berger continues. "They belonged *there* and *here*. . . . They were mortal and immortal. An animal's blood flowed like human blood, but its species was undying. . . . This, maybe the first existential dualism, was reflected in the treatment of animals. They were subjected *and* worshipped, bred *and* sacrificed.

"Today," Berger continues, "the vestiges of this dualism remain among those who live intimately with, and depend upon, animals. A peasant becomes fond of his pig and is glad to salt away its pork. What is significant, and is so difficult for the urban stranger to understand, is that the two statements in that sentence are connected by an *and* not by a *but*."

Animals provoked, and still provoke, some of humans' first questions. Questions that, today, most of us would rather not have to grapple with. By refusing to grapple, by living in the land of *buts* and forsaking the *ands,* we can easily come to believe we've absolved ourselves from ever having to confront those difficult questions.

An *and,* not a *but.* That is what that split pig's brain in my palm felt like.

"The test of first-rate intelligence is the ability to hold two opposed ideas in mind at the same time and still retain the ability to function," F. Scott Fitzgerald wrote.

It seemed to me, standing there with two halves of a pig brain cupped in my palm, that we are often terrible at this kind of first-rate intelligence, that, in fact, so much of what we do is in the service of keeping opposing ideas at bay inside ourselves. Isn't this what we're doing when we eat meat without taking part in the process that brings it to our tables, without ever being required to stare back at the animal that made that meat possible? Did we not grow our industrial food complex precisely so that we didn't have to simultaneously become fond of our pig and be glad to salt it, too?

SEVENTEEN

As my first days turned into weeks, the air in Gascony turned dry and hot. Sunflower fields glowed bright yellow. Kate sent me on frequent bike rides along the canal to hunt for wild elderberries to harvest for jams and syrups and liqueur. Each market visit brought new spoils. The tiniest of strawberries, with big, loud flavor. Shiny red cherries. Honeyed melons. When I wasn't in the Chapolards' *salle de découpe* or working at the market, I weeded Kate's garden, which proved challenging to keep up with. Each day, I collected fresh chicken and duck eggs for breakfast, harvested greens and nasturtium flowers for dinner. I dug a big hole in the duck yard and lined it with plastic to make a pond for Kate's ducks. Two more sets of students came and went. This time they seemed to appreciate what Kate had to show them—and they, like me, never wanted to leave. Kate taught me how to make flaky, buttery scones my mother would approve of. And she brought me to meet Monsieur Gros, whom my friend Robert Reynolds had talked about as if he were a god.

A renowned eau-de-vie producer in the region, Monsieur Gros turned out a tiny production, but he had been making small-batch eau-de-vie ("water of life") for several decades. This described just about every producer I met in Gascony: renowned, tiny production, decades of knowledge. Monsieur Gros made *eau-de-vie de fruits,* an un-aged brandy made by fermenting and twice distilling ripe fruits like pears or apples. But he was also known for another type of eau-de-vie. The story goes that one day, Monsieur Gros and his wife stood in the doorway of their stone distillery, looking out at all their grapevines, and, as if for the first time, they saw the wild *chèvrefeuille,* or honey-suckle, growing between the vines. "We should drink that," his wife said. And so they began transforming all those pink and purple flowers into liquid in the form of honeysuckle eau-de-vie, a most quintessential form of Gascon alchemy. Just writing this makes me smile. I bought as many bottles as I could fit in my suitcase.

After our visit to Monsieur Gros, Kate took me to the prune farm near her house, where I tasted prune eau-de-vie, as well as candied prunes, chocolate-covered prunes, and just plain prunes that tasted nothing like what I'd had back home. They were full of sugar, but with a touch of musk and a lemony acidity. On Fruitway Road, the Italian plum orchard we lived across from had once been nationally renowned for the prunes its plums were turned into, but by the time I'd entered elementary school, the orchard's owners had abandoned it. What was it that had allowed Gascony's plum orchards to thrive and Alvadore's to fail? It was as if at some point Alvadore's inhabitants had just stopped seeing all the potential around them. We couldn't see the honeysuckle. And even if we had been able to see it, we were no longer in possession of the proper stories, stories that would tell us what to do with it, stories that would reawaken our own spirit of *débrouillardise.*

MONSIEUR GROS'S honeysuckle story reminded me once again of how very difficult it can be to truly see. How many times had he and his wife gone out to look at their vines and never really noticed the honey-suckle and its potential? How many times had I bought meat and never really seen what I was buying?

Annie Dillard wrote that seeing is most often a matter of verbaliza-tion, or expression in some form or another. "Unless I call my attention to what passes before my eyes," she wrote, "I simply won't see it." Writ-ing, for so much of my life, had done this for me.

"But there is another kind of seeing," Dillard wrote, "that involves a letting go. When I see this way I sway transfixed and emptied. The difference between the two ways of seeing is the difference between walking with and without a camera. When I walk with a camera I walk from shot to shot, reading the light on a calibrated meter. When I walk without a camera, my own shutter opens, and the moment's light prints on my own silver gut. When I see this second way I am above all an unscrupulous observer."

This is what choosing to stand on the other side of a pig did to me, the side opposite from where I'd stood for thirty-two years, the side that required me to pick up a knife, to read the road map of the pig's anat-omy, to endow each muscle with its own story.

As time beat on in France, I wished to rely solely on each moment's light, which I imagined to be the bright, obtrusive kind, imprinting it-self on my neglected silver gut. This made me an increasingly poor communicator with the people I'd left back home, however. At first I'd written diligently to Will, and to my family and close friends—I'd even tried my hand at writing a few Facebook and blog posts—but as my time working in the cutting room and at the market pushed me more and more into the present, I became less interested in describing my experience to anyone. And while I sensed that this would reinforce the distance that had likely formed between me and those I loved back home, I didn't really care. It felt reckless, but that silver light kept catching me in its particulate wild glow.

And so I forgot my friend Sonya's birthday. I forgot that my friend Jill was about to have her baby. Had I forgotten Mother's Day and Father's Day, too? What else had I forgotten? Who else?

Will. I had forgotten to call Will on his birthday, and I didn't realize it until two weeks later.

"I'm so sorry," I said to a mostly frozen, very delayed image of him over Skype. "I'm just so . . . caught up here."

Will assured me, over and over again, that it was fine. He didn't care. He understood. He employed all the right phrases—"I want to give you the space you need"—that one should when they wish to sound confident and unselfish. But in his voice I sensed the tremble of a growing hurt.

It wasn't that I didn't care about his birthday, but somehow it hadn't remained a priority while I was in France. I was too busy opening my eyes to *see*. And this, I realized, was probably as good a reason as any *not* to be with someone, at least in the way I had so hastily chosen to *be* with Will. *I love you* AND *I want to be alone. I love you* BUT *I want to be alone.* There is danger in seeing, too. You come to unscrupulous conclusions. You make hard decisions. You will always suffer consequences.

That night I dreamed that I'd returned to Will's house. I was disoriented in the dream and felt as if I were losing my eyesight. Things in the house had been moved around. I couldn't make my way around all the furniture. I didn't recognize the place at all. And my teeth, how they ached.

EIGHTEEN

Two holes had torn in the fabric and I could not stop looking through them. The fabric was translucent, made of loosely woven black mesh, nailed from the top of the barn doorway to keep out the flies or dust or detritus, perhaps, once, long ago. The elements had worn the fabric, torn it away from the doorway, so it was tattered, flapping in the wind, with two fist-size holes in it. Through the holes, the blue sky, a wispy Gascon cloud, a fir tree. This is what I took hundreds of pictures of at Jehanne's *ferme auberge*, where she showed Jonathan and Kate and me how she raised fattened ducks and geese the old-fashioned way, with a beat-up metal funnel and some grain.

I took the pictures of the holes in the fabric with a new camera that Will had given me before I left for France. At the end of our last, awkward phone call, he'd asked me to send photos meant just for him. The photos I'd sent in bulk e-mails to friends and family hadn't done it for him. "Special photos," he said, "with the camera I gave you."

It was an innocent enough request—maybe even romantic in its own way. And besides, I owed him for having missed his birthday. But it made me impatient. I didn't really want to take photos for Will, or anyone else. I didn't want to feel obligated to anyone, even this man I had fallen in love with. This man I was not sure I would feel the same way about when I returned home.

Give me a little space, I wanted to say, *a little time, some room to grow.* But I'd moved in and then left in a hurry and was now out of reach and unwilling to share. I could understand his worry. And then all I had to show him was picture after picture of those two holes interrupting that dark veil of material to reveal the blue sky.

WHEN WE ARRIVED, Jehanne showed us the open-air barn and pasture that lay beyond, where her ducks and geese sauntered and waddled in the open yard and dipped their beaks in troughs of grain. The gray and white down of the geese resembled that of the Canadian variety we grew up being chased by, the ones our neighbors on Fruitway Road kept for fun, presumably, although I never understood why, because they were loud and mean and provided little in the way of companionship. Jehanne's geese had much shorter necks and legs, and they seemed a good deal calmer.

Underneath the barn's roof, and closed in by temporary plastic netting, fifty or so ducklings lay in the shade on the ground, their new black and yellow feathers fuzzy and soft, sticking out in every direction, alive to the new air around them.

"These ducks are young," Jehanne said, speaking pragmatically. "We want to protect them from the elements in here for a few more weeks." She noted that they could easily move around and probe the dirt with their beaks. She did not like to put them in cages. "Once they're older, we'll rotate them on our various pastures."

"For the last two weeks of their sixteen or so weeks of life, we do the *gavage*," Jehanne explained. *Gavage* is the French term for what we in America have dubbed "force-feeding," although the direct translation is "to gorge." "We do *gavage* the old way."

Kate explained that industrial foie gras birds are typically raised in individual cages or cramped barns and that they are often force-fed for a shorter period of time with larger quantities of food at each feeding.

Jehanne calmly corralled one of the older geese into an enclosed area of the barn. She sat on the ground and coaxed the bird between her legs to keep it from flapping its wings. She then grabbed what looked like a weathered metal funnel with a slender, slightly longer-than-normal stem, used it to scoop up a half-cup or so of wet-looking corn from a bucket, squeezed the sides of the goose's beak to open its mouth, and gently inserted the stem of the funnel a few inches down the bird's throat. Jehanne pressed on a loop of metal at the middle of the funnel to release the corn, several kernels at a time, while she massaged the goose's neck.

"She's inserting it into the bird's crop," Kate explained, "which sits at the base of the esophagus. That's where the corn ends up. It doesn't go directly into the stomach. Over time, the bird releases the food from the crop into the gizzard and then the stomach."

After about fifteen seconds, Jehanne removed the funnel and the bird waddled off.

"That's it?" I asked. The goose seemed completely unfazed, although of course I could not be sure what the bird felt, and were I a different kind of person in need of a different sort of narrative, I could lay an entirely different story onto what I saw.

I tried to imagine what *gavage* might feel like if it were done to me. I pictured myself gagging. I would likely have a hard time breathing, and the stem of the funnel would surely hurt my throat. But I was working off the assumption that the goose and I had the same anatomy.

Kate explained that ducks and geese don't have teeth, and they normally swallow their food whole, storing it first in their crop. They also swallow rocks and pebbles, which end up in the gizzard. Once the food moves from the crop to the gizzard, those rocks and pebbles grind up the food before it's digested. Because of this particular system, the bird's esophagus is actually quite flexible and resilient, unlike a human's. Ducks don't have a gag reflex, either, because their breathing and feeding apparatuses are separate. My natural empathetic assumptions about what it might feel like for the goose were entirely off base.

I imagined that the goose felt at least a little discomfort, but, judging by its response, the process didn't seem to cause physical pain. Discomfort and pain are two very different things, after all. I experience discomfort almost every day, when a bug bite itches or when I get a splinter or a seat belt becomes too tight around my chest, but pain is something I generally try to avoid and never want to inflict on others. Is causing discomfort acceptable to some people but not to others? Are discomfort and pain the same in some people's minds? Discomfort was part of life here on Jehanne's farm. But she seemed to be saying that pain was where they drew their line.

I could see how much work it was to do it Jehanne's way, how it might be tempting to mechanize the whole thing, to keep all the birds in cages so that you didn't have to hold the bird in your lap and be reminded that it was a living creature. But this seemed like the only respectable way to do it, if you were going to do it. If I could not find anyone back in the States who raised foie gras like Jehanne did, I wasn't sure I wanted to eat foie gras at all.

JEHANNE WALKED US around the corner of the barn to their small processing plant. Outside, a tall man with a shaved head and a trim black goatee was in the process of killing a batch of Jehanne's ducks. Kate explained that the man was Jehanne's informally adopted son, a once wayward kid she'd taken under her wing long ago. He wore spotless white pants, a white butcher's smock, and a white rubber apron that hung down to his ankles, and he stood in front of a contraption that looked like four traffic cones turned upside down and tied to a tripod. Beneath it he'd laid a burlap rag on the gravel to catch the blood, a gesture that seemed more ceremonial than practical.

Two duck feet stuck out of the top of each cone. Jehanne explained that before killing each duck, they placed it upside down in the cone to keep it from flapping its wings and to calm it. The opening at the small end of each cone was just big enough that the duck's head and neck could be gently pulled through so that it wasn't scrunched up.

The man opened one of the duck's beaks and quickly stuck the tip of a knife through the thin top palate of the bird's mouth and into the brain. After pulling the knife back out, he stuck it into the duck's neck, directly into its carotid artery, to bleed it. This was more or less the same procedure I'd witnessed at the abattoir two Sundays earlier. Scramble the brain-body communication so it can't feel pain. Then drain the blood. Do it quick. Don't mess it up.

I thought of the stories my dad told me about killing chickens as a child, how his grandma chopped the head off with a hatchet, and the headless chicken chased him around the yard. Jehanne's birds were not struggling or running in the instant of their death. They were still and calm when the lights went out.

Once the duck bled out completely, the man passed each bird through an opening in the wall of the building and Jehanne motioned for us to follow her inside. A heavy dusting of feathers covered the floor. Red-tinted water flowed into a drain.

Two women and a man worked to rid the ducks of all their feathers. First the man dipped a duck into a vat of hot water. "To loosen the feathers," Jehanne said. Then, he held the duck above a contraption that looked a bit like a pinball machine with a system of rapidly moving rollers and rubber fingers that pulled the majority of the feathers away from the carcass, leaving the skin intact. When he was done, the two women hung the birds from the ceiling by their feet and removed the rest using their fingers, a knife, and a wet towel. Before passing these cleaned duck carcasses through another window, they cut through the skin at the base of the neck and pulled the neck skin down toward the ground so that it hung loosely from the carcass but remained attached.

"We use the neck skin to make *chou farci*," Kate said. "You stuff it with foie gras or with ground duck meat. Jehanne serves it to her guests here and cans and sells it to customers at the market."

Kate pointed to the creamy yellow fat underneath the skin. "The birds have so much fat on them that when you render the fat from just one duck, you have enough to properly confit both legs."

Confit refers to the process of cooking any ingredient, but usually meat, at a low temperature for a long time in copious amounts of fat— enough fat to cover the ingredient, in fact. The fat is usually from the same animal the meat came from. Someone had long ago figured out that if you feed the birds a lot of corn, the bird will have enough fat for you to confit the legs in, a process that turns the meat succulent and tender. And if you place the confited legs in a jar and cover them with rendered fat before sealing it, the meat will be preserved until you open it months—even years—later. Without that fat, all manner of traditional duck preservation methods would not exist.

In the next room, a young woman butchered the ducks, still warm. First she put her entire hand inside the cavity of the bird and pulled out the organs, separating out heart, lungs, spleen, and liver, which, by far the largest of the organs, was the color of burnt caramel, with a bit of creamy yellow mixed in. The liver appeared to have taken up the majority of the inner cavity of the bird. Kate told us the duck carcasses weighed eleven to twelve pounds and that the liver accounted for 10 percent of that weight. Would that enlarged liver have been painful? Uncomfortable? Do we have fatty livers, I wondered, given the way many of us eat? Possibly, but we go about our lives anyway.

With just a few swipes of her knife, the woman magically pulled the entire main skeleton, including the rib cage, spine, and neck, out in one piece and tossed it into a big pot. Two breasts, two wings, two thighs, and two legs remained, all held together by a sheath of duck skin and fat. The resulting patchwork looked a bit like a deep-yellow and dark-red Rorschach print of muscle, fat, and skin.

"They cut it this way," Kate explained, "so they can control exactly how much skin and fat stays on each piece. They use every part of this bird for a different product, and the amount of skin and fat on each is very important."

Jehanne explained that she preferred to hot-butcher her birds so that every part was as fresh as possible for salting or cooking, and then she shooed us out of the cutting room. As we took off our hairnets and white coats, I asked Jehanne if I could have the recipe for her foie-gras-stuffed duck prosciutto.

"I don't think the recipe will work anywhere but here," she said.

I laughed.

"She's not joking," Kate said. "It will be difficult to find foie gras in the States as fresh as this."

Two weeks later, Kate took some of her students and me back to Jehanne's *ferme auberge* for lunch. French, German, Swiss, American, and Spanish tourists filled the dining room, which felt a bit like a ski lodge, with its vaulted ceilings and long, wooden communal tables. For the first course, we were served flutes of sparkling wine and tastes of Jehanne's salted, dried duck prosciutto. For each course, Jehanne stood up on a stool to tell the entire room what we were about to eat, and as the courses progressed, we cheered every time she did so.

For the second course, she brought out a terrine of silky, peppery garlic soup for each table. Jehanne's broth was sturdy and soothing, made rich by way of slow-cooked, roasted duck bones and a lot of minced garlic from her garden. Water. Bones. Garlic. Salt. Pepper. Plus, an egg or two whisked in to achieve that silkiness. That was it, and yet it very quickly fortified us, inspiring us to sit up a little straighter.

Next, slices of toasted bread, which Jehanne instructed us to rub with a clove of garlic before slathering her award-winning rillettes onto it. Jehanne made her rillettes by picking and shredding the meat from the bones she used to make her soup broth and whipping that meat with rendered duck fat. She seasoned her spreadable, rough pâté with just salt and pepper, rendering its flavor earthy transformative. Like confit, rillettes was a recipe born out of resourcefulness, out of a desire to waste nothing, to use every single part of the animal whose life we took for food. Jehanne's rillettes possessed the flavor and texture of *débrouillardise*.

A platter of duck confit with lentils and bitter garden greens arrived.

I'd eaten confit many times at restaurants back in the States, but this duck looked and tasted different. The meat was dark red, almost purple, as opposed to a light pink. It possessed structure and bite, its texture sturdy, even if at the same time it was tender and falling off the bone.

"It's the fat that makes this duck taste Gascon," Kate said. "You'll never find this back home unless you raise the duck like Jehanne does."

I nearly protested. Of course I could find a Jehanne back home. I lived in Portland, after all. If anyone was doing what Jehanne was doing in the United States, I would find her there. But I didn't really know who or what I would find back home. I hadn't known what I should look for until I came to Gascony.

NINETEEN

Learning a new skill, especially as an adult, requires you to forget what you know, or what you think you know, on a regular basis.

By my fourth week working with the Chapolards, I felt I'd watched them break down enough sides that I could do it myself without too much trouble. I even sat down with Christiane before lunch at her house one day to show her what I knew, drawing for her a rough sketch of a side of pig and then, without much help from Christiane, labeling, in French and English, each of the primals and subprimals. I was also able to draw where, more or less, each primal should be cut and indicate which bones I would subsequently have to remove. I had this. I'd put on my reporter's cap and gleaned all the information I could. I'd done all my homework.

A few days later, Bruno and Dominique let both Jonathan and me cut up our own sides of pork in the *salle de découpe*. Kate, our proud mother, came to take pictures.

In one of the pictures, I'm holding a meat saw in my right hand. Dominique had offered it to me to assist in removing the back leg, since I was running into some trouble getting my knife through a particularly tricky bone. To make the saw go, I've stepped one of my back legs behind me for leverage, as if at the starting line of a race, grinding my foot into the ground. The front of my body leans at a forty-five-degree angle toward the metal table. My left hand rests on the pig's rear end and I'm mustering all the power I can to send energy into the saw, which I'm pushing away from my body with all my might.

Dominique watched me until I finally got through the sacrum. "Camees, when you work on the pig, stand up straight, breathe, and smile."

He said something to Kate.

"If you angle the saw correctly and the teeth are facing the right direction," Kate translated, "you shouldn't have to work that hard."

In theory, I understood which bones to cut through, which muscles to separate from which muscles. But when I employed my body to do the job, none of that really mattered. My body knew nothing. And my body was the most important part of completing this task in front of me. I needed my body to learn, too. I needed to remember where my arm should be, how my elbow needed to move, how to keep my wrist and back straight. And so my body felt like a new and awkward tool, as if I'd moved my hands in front of my eyes and realized for the first time that they were my hands to move and not someone else's.

Dominique moved my ham out of the way and asked me what came next.

"I'm going to remove the trotter and hock from the shoulder," I said. He nodded in approval.

I set the saw down, picked up a short, flexible boning knife, sliced through the small amount of skin, fat, and muscle that surrounded the ankle joint, and began bending my knife in between the joint's awkward curves, searching for the ligament that held the two knuckle-shaped bones together. Then a quick snap, which I could feel down the length of the knife and into my hand. I'd severed the ligament cleanly and could now separate the trotter from the hock.

"*Formidable,*" Dominique said, his face brightening.

The act of butchery is, if nothing else, an immediate one requiring you to locate your own body, but also the pig's body, in the present tense. As I located each ligament, each line of fascia, each muscle and tendon, I began to see this pig's individual, particular road map. Sure, every pig has a front leg and a back leg, a belly and a loin, but I was beginning to see tiny differences in bone and muscle development between each pig—the tributaries, the barely discernible dirt roads. Being able to see the subtleties of each pig's individual road map seemed essential, because each pig was different. Were I working in an industrialized meat-processing facility, butchering hundreds of pig carcasses every day, I imagine there simply wouldn't be enough time to be able to read each animal in this way.

Yet it was precisely through these readings that the Chapolards could tell whether the quality of the wheat and barley they grew for their pigs was sufficient. Were their pigs getting enough exercise? Were they sick or healthy? Fat or lean? Were they slaughtering them at the right age? Had the breeding lines remained pure? In order to ascertain this information, they needed their own bodies to be in close proximity to those of the animals.

What might it be like if we all lived in such close proximity to the animals we ate? If we had to perform, or at least be witness to, the work of these saws and knives and cleavers in order to put meat on our own tables? How much meat might we eat then? How much might we be willing to pay for someone else to do the close reading for us if we understood the difficult paradox it required?

WITH THE BACK and front legs removed, I turned my attention to the loin and belly. Dominique gave me the option of turning the loin into pork chops or roasts.

I have, since childhood, been extremely biased against pork chops. It was the only "fancy" cut I remember eating as a kid—or at least the only one more expensive than ground meat or stew cubes—so whenever my mom cooked pork chops, it often felt like a special occasion. But she'd cook them for so long that I'd sit at the table for what felt like hours, chewing. I found them to be dry and unpleasant, no matter how delicious a sauce my mother made, and so, early on in life, I concluded that all pork chops were bad.

If you ask my mom why she overcooks pork chops, she'll tell you that she simply doesn't like "bloody meat." If I explain to her that there's no blood in meat, because the animal is bled after slaughter and well before we eat it, she'll say she simply finds any hint of rawness "disgusting."

Her answers have always struck me as irrational, but she is most definitely not alone in her reaction. In fact, her tough pork chops can probably be credited to a tiny roundworm called trichina. Discovered in the mid-1800s, trichinae can infect carnivorous and omnivorous animals, from domestic pigs to wild cougars. If humans eat undercooked meat from infected animals, they, too, can contract trichinosis, which can result in a whole host of unpleasant symptoms and even death. The major reason that pigs used to get trichinosis? They were fed uncooked table scraps and animal carcasses that were already infected, or they came into contact with the carcasses of other infected wild animals. Industrial farming, which brought pigs indoors, mostly eliminated the latter. But it wasn't until the 1980s that feeding uncooked food scraps and waste to domestic pigs became illegal, so the only way to ensure that we didn't get trichinosis was to cook our pork really well. To be safe, the USDA suggests cooking ground pork and organ and variety meats until the internal temperature reaches 160 degrees Fahrenheit, and raw steaks, chops, and roasts to a minimum of 145 degrees, though trichinae can't live for more than a minute in 140-degree temperatures.

In other words, maybe my mom's disgust at the thought of "bloody meat" is really about a passed-down, mostly subconscious avoidance of sickness.

However, if indoor and outdoor pigs are managed properly, and their exposure to potential carriers is eliminated, trichinosis is not a problem. And the thing is, pork chops—especially chops from the young, lean pigs we raise in the States—really do become dry, chewy, and hard if you cook them to that temperature. This, I would eventually learn, is because the loin muscle that pork chops come from is a structural support muscle, its job being primarily to hold up the pig's body. It's not used for locomotion in the same way that shoulder or leg muscles—which must move in a slow, sustained manner almost constantly—are. Thus, the fibers of the loin muscle are bundled quite differently than locomotive muscles, such that the loin contains much less connective tissue—what we often refer to as gristle—within the muscle. As a result, structural support muscles like the loin and the tenderloin are typically more tender to begin with, requiring a faster, hotter cooking method, whereas any locomotive muscle or cut containing that gristle will require a longer and slower cooking time in order to render that tough gristle more tender. Cooked right, both kinds of muscles can be tender, but cooked wrong, they will be tough. Too much high heat for too long, for instance, will cause a loin muscle's proteins to bind tightly together and never let go, hence all my chewing at the dinner table.

Despite knowing that pork chops, when cooked right, could be delicious, I chose to cut the loin into roasts. Plus, I wanted to try my hand at what I had seen Dominique do that first week: *opening the pig like a book*.

With the tip of my knife, I gently traced around the tip of each rib until I could get underneath the rib cage and begin carving long lines with my knife, to release the fence of ribs. I didn't yet understand how to create a smooth line between rib and belly, so I covered the belly with hack marks that I'd have to go back and smooth out with my knife later.

I knew in my brain that the loin muscle curved up into the spine and that I had to be careful to curve my knife along with it, but my hand and body didn't know this yet, so I shaved about a quarter of the loin off along with the column of vertebrae I removed. I completed the cuts I was supposed to, but in rough and imprecise ways. My bones had meat on them and they were not supposed to. The loin muscle was missing its top side.

Kate and Dominique were forgiving. *We can always make saucisse,* they said.

After Jonathan and I finished cutting our sides of pork, it was quitting time. My hands and legs and arms felt sore. I'd mostly white-knuckled my way through and worked way too hard, but I'd completed my task. I'd not gotten lost too many times, and next time, I thought, maybe my body would remember all the tricky curves.

SINCE JULY 4 was right around the corner, Kate and Jonathan and I planned to throw a party and roast a suckling pig for it. We'd make barbecue sauce, baked beans, and potato salad. We'd invite all our American and French friends to play boules and drink cold rosé with us.

There's another picture Kate took that day in the cutting room, of

Dominique and me posing with a freshly harvested suckling pig that we'd bought from the Chapolards. I'm not fond of that picture now. In it, we're clearly calling attention to how small the pig is, looking upon the pig as we would a human baby, in a joking sort of way. While Kate took the picture, Dominique told me he didn't like suckling pigs because they had no meat on them and tasted like nothing.

"Tourists like it, though."

On the way to Kate's, I asked her to explain.

"You slaughter your pigs really young in America, at about six months. The Chapolards slaughter their pigs at twelve months, sometimes more. The beef I buy in Gascony comes from cows that are usually four to seven years old. Your beef comes from animals that are closer to two years, or even less. You guys like your meat to taste mild. We like meat with character, with well-developed fat and connective tissue, which are what give meat flavor."

I asked her if the age of the animal also explained the rich color of the Chapolards' pork, which was so much redder than the pale pink pork I normally ate back home.

"Exactly," she said. "The redder the pork, the older it is, typically; the more the pig has moved around, the more concentrated flavor it's going to have." Pork in America largely came from young factory-farmed pigs that barely moved around during their lifetimes.

Americans eat a whopping 265 pounds of meat per capita per year, second only to Australia. Perhaps this lack of flavor and texture in American meat—not to mention its low price—explained why a sixteen-ounce rib eye had become a central point of pride on our dinner plates. Did we have to eat that entire sixteen-ounce rib eye just to feel sated? How would our meat eating differ if we ate older animals, animals allowed to move around, fed a diverse diet—animals with complex flavor and texture, which necessarily cost more because they had to be fed for a longer period of time? We'd probably eat a lot less meat, like my friends in Gascony did.

FOR OUR INDEPENDENCE DAY PARTY, I made pimento cheese with French Mimolette instead of American cheddar, which I could not find—and smoked a pork shoulder for pulled pork, to make up for the fact that the suckling pig had such a small amount of meat on it.

I'd lost all my enthusiasm for roasting the suckling pig, which, now that Dominique and Kate had schooled me, felt like a pointless waste. In fact, I decided this would be the last suckling pig I'd ever eat or roast.

That evening, while we Americans proudly, if a touch ironically, drank cheap beer from cans and piled our plates high with food, our French friends picked at their food.

"What is this orange stuff?" Dominique asked me, sniffing the pimento cheese suspiciously.

Our French guests didn't eat much of the suckling pig, and they asked why our baked beans were so sweet.

"We add brown sugar and molasses," I told them.

"Why would you do that?" Kate's friend Vètou, a surly woman with a strong patois who had taught Kate much of what she knew about Gascon cooking, asked us. "You can't taste the beans."

TWENTY

A week after our Independence Day party, my friend Eugenie arrived from Portland to take photos for Kate's new Web site in exchange for a week or two of room and board. Bill, a twentysomething ex–professional skateboarder and recent culinary school graduate, also showed up. After Bill and Eugenie settled in and we all found our new rhythm together, Kate challenged all of us to preserve one whole pig in one hundred pressure-cooked mason jars, give or take. We were allowed to preserve the pig in ways other than the mason jars, but she needed to be able to store just about all of it in her piggery without refrigeration. The entire pig would feed Kate and her students and dinner guests for the next year, and she didn't have a lot of freezer space, so we would need to use what we learned over the past five weeks to figure it out. Jonathan and I would each get a side of pork from the Chapolards to break down on Kate's kitchen counter, and then the five of us would attempt to turn everything into food.

This would be my third time breaking down a side of pig by myself—this time without the help of the Chapolards, although Kate would be there to guide us.

As I separated each of the primals, I contemplated the role each muscle played in the pig's everyday movements—movements that made the meat look as red as it did, or as pale, or as filled with fat, or as tightly or loosely grained—so that once I'd taken it all apart, I could still remember where each muscle had come from and apply my knowledge of each muscle's story in the kitchen.

My internal dialogue with the muscles in front of me went something like this: "Hello, coppa, it's nice to meet you. You live at the front end of the loin, just around the corner from the jowl. Back home in the States, you make up part of the Boston butt and you live upstairs from the 'picnic.' You work hard every day while the pig roots around and shakes its head *yes* and *no* and *maybe* and eats and drinks water. And because you work so hard, in the same way, over and over, every day, you have well-developed muscles with a lot of connective fat and tissue in between them and even more flavor. You will be tough to get to know until we render you soft and tender. You can stand a little heat over a long period of time. 'Slow and low' is your motto. But you also don't mind sitting in a whole lot of salt before being hung out to dry."

To get better at this, to become an expert reader of the pig's road map, I would need to repeat this process again and again when I got home, but I couldn't very well afford to buy myself that many sides of pig to work on with my meager unemployment checks. And even if I could, I was no longer sure I wanted that much meat in my life.

As Jonathan and I parceled out our sides, we began to make piles just as the Chapolards had shown us. One for sausage meat. One for all the hams we would salt and smoke. One for the ham meat we'd use to make *jambon de Tonneins,* a highly herbaceous Gascon version of pulled pork, but made from the back leg instead of the shoulder. On the stove we had a pot going for some of the extra fat, which we would render and use to make rillettes or *gratton.* We started another pile for pâté, which we planned to seal in jars. We would make gelatinous stock out of the bones and skin, and Kate could use that for canning beans or tomatoes from her garden. We kept the bellies whole to make *ventrèche,* which we would smoke and then hang, along with the hams, in Kate's piggery, the coolest room of her house. Any other belly scraps would be added to the rillettes or pâté piles.

Everything we planned to make would be used as accents to Kate's everlasting meal. Just small amounts of highly concentrated, flavorful meats, preserved in fat or liquid or salt, with the help of heat, smoke, or the open air.

It took us each a little over an hour to cut up our pigs, and we were tired afterward, but we didn't have enough refrigerators to store all the meat for the next day, so we needed to start cooking and salting right away. This process normally would have been done in the fall, when it was cooler, but for the sake of our education, we were doing it in July.

We salted the bellies, which would have to fit into Kate's small refrigerator for a day or two before smoking. We salted our hams and the tenderloin, all of which would eventually be smoked and/or hung to dry. We talked about salting some of the loin to turn it into Spanish-style *lomo,* but then Kate suggested we make a small batch of paupiettes and preserve those in rendered fat in mason jars.

Kate pulled out an antique, hand-cranked meat grinder that we attached by way of a vise to the kitchen table, and I began to grind meat and fat for sausage and pâté. But it was too warm, so it smeared out of the grinder like a paste, as opposed to neatly separated coils.

"This is why we do this when it's cold outside," Kate said. "We'll have to put the meat in the freezer for a while." Thank goodness for freezers, even small ones like Kate's.

We put smaller pieces of fresh ham muscle in a Dutch oven on the stove and covered it with white wine and bay leaves, a smattering of peppercorns, and lots of olive oil to make the *jambons de Tonneins*. And we began cooking our stock and rillettes.

Everything cooked on low heat on the stove or in the oven overnight, and in the morning we commenced to stuffing jars with all of our recipes. We filled two very tall canning pots with water, stacked the jars inside, and set the pots over portable burners outside. While we waited for the cans to seal, we began stuffing sausages—simple ones, just like the Chapolards made. Salt. Pepper. Pork. Fat. The sausage-stuffing attachment for her antique grinder was a bit rickety, but we managed.

By lunchtime, Bill, Eugenie, Jonathan, and I were flagging. We took a break by the boules court, sprayed one another with the hose to cool down, and drank lukewarm beer with ice in it. But after a few minutes, Kate found us.

"There's no time for rest now, guys. It's hot out. I'm paying for this meat. I'm the one who has to eat it. I need you to help me get this done."

Our intention had been to take a short break and then return to work, but I felt guilty nonetheless. Even though we understood the limitations of Kate's refrigeration, the whole thing felt a bit like an abstract learning exercise to me. I was still a tourist. For Kate, this was her life—not a lifestyle—and it was how she was going to eat for the next year. We were all going to go home soon to America, where meat was everywhere, easy to find, easy to buy, on everyone's plates—where, if we chose, we wouldn't have to think about any of it ever again. Not that this was my plan. In fact, I couldn't wrap my head around going back to the way things were.

I had the sense that finding the Chapolards of America, acquiring the kind of meat they raised, ensuring that I, too, had a pantry full of *jambon* and rillettes, was going to require just as much work as it did here in Gascony, if not more. Living and eating and cooking like Kate meant you had to work. You had to be a rapt and active participant in nearly every step of the process of getting food to your table. Sure, it was hard, but I liked to think I was the kind of person for whom this sort of work was fulfilling. I hadn't known I was that kind of person until I went to Gascony. But I was. I still am.

AT THE END of two long days of pig processing, we all pitched in to prepare what Kate called leftovers but what felt to me, as usual, like an extravagant feast. We toasted bread over a hot fire we built in her outdoor oven and spread our pork rillettes over it. The Swiss chard in Kate's garden was about to bolt, so Kate taught me how to make a *tourte aux blettes*, a sweet and savory Provençal tart made from currants, Swiss chard, eggs, and pine nuts. We added chopped prunes to make it more Gascon. We gathered greens from her garden for a salad, pulled out whatever stinky cheese in Kate's cheese stash needed eating, and sat outside devouring our leftovers and washing it all down with tart rosé dispensed from a box that we'd bought from a local wine cooperative earlier in the week.

Afterward, we all lazed about outside Kate's outdoor trailer office and hooked into her Wi-Fi so we could each catch up on the rest of the world, which I'd ignored for quite some time. Kate sat working on her computer in her trailer. Bacon lay on the floor by her side. Eugenie lay in a hammock next to Jonathan, who'd taken a much quicker shine to her than he had to me. Bill poked around in Kate's overgrown *potager* nearby.

I was skimming through articles online when a piece in *The New York Times* caught my eye. "Young Idols with Cleavers Rule the Stage," the headline read.

The article talked about a couple, Joshua and Jessica Applestone, and their Fleishers butcher shop, in upstate New York, how they were sourcing animals from local, humane farmers and attempting to sell the whole animal to their customers. A chef named Ryan Farr had begun holding butchery demonstrations in California bars and was charging good money for them. Avedano's, a female-run shop that had just opened up in San Francisco, had begun offering pig butchery classes, and a guy named Tom Mylan, who headed up the meat program at Marlow & Daughters, in Brooklyn, had also started doing public butchery demonstrations. All of these people had several things in common: They were trying to change how we thought about meat in America by going back to a more traditional way of doing things. They were dedicated to whole-animal butchery and utilization, and to throwing open at least some of the closed doors of our meat industry so that consumers could start to understand where meat came from. And they were young—the new thirtysomething young, like me—and many had once succeeded at other, completely unrelated careers and then decided, seemingly around the same time, to become butchers.

I'd had no idea.

"I guess you're going home at exactly the right time," Kate said.

But I didn't like what I read:

"If chefs were rock stars, they would be arena bands, playing hard and loud with thousands cheering.

"Farmers, who gently coax food from the earth, are more like folk singers, less flashy and more introspective.

"Now there is a new kind of star on the food scene: young butchers. With their swinging scabbards, muscled forearms, and constant proximity to flesh, butchers have the raw, emotional appeal of an indie band. They turn death into life, in the form of a really good skirt steak."

I could not help but roll my eyes. Swinging scabbards? Muscled forearms?

It was so very American the way the *Times* had distanced these new butchers—every one of whom I wanted to meet—from reality by likening the act of turning animals into food to performance.

"Seriously?" I said, turning to the others. "An indie band?"

Little did I know that the media would soon try to do the very same thing to me.

TWENTY-ONE

A week before I returned to Oregon, Eugenie and I rented a car and meandered our way through northern Spain for a week. We ate salty ham and briny olives every day, washing it all down with sweet, herbaceous vermouth. We swam naked in the ocean off the Costa Brava and sunbathed on the same beaches that Salvador Dalí wandered as a boy. We stocked up on red-and-gold tins of piment d'Espelette in the hills of Spain's Basque country and stuffed ourselves with *pintxos* in San Sebastián. Our time in Spain felt luxurious, maybe even undeserved, and eventually the idleness of it all—not to mention how close I was to maxing out my credit card—begin to make me anxious, though I tried my best to fight it, to allow myself this one last break from real life back home. *Real life*. Whatever that meant.

After dropping Eugenie off at the airport, I returned to Kate's for a few more days. She'd promised to help me brainstorm how I might take what I'd learned in Gascony back home and do something of value with it, as Christiane had urged me to. I dreamed of opening up my own portable butcher shop like the Chapolards', and Kate and I were excited about the name we'd come up with: La Belle Bouche. *Bouche* is short for *boucherie*, or "butchery," but it is also the word for "mouth." *Belle* means "beautiful," and we felt it added a touch of the feminine to a largely male realm. I loved it. It felt defiant and welcoming in the same breath. But when I shared the name with a few friends back home, they separately reminded me of that famously creepy "purty mouth" line from the movie *Deliverance*. Leave it to Americans to turn a lovely French name for a butcher shop into a reference to backwoods Appalachian sodomy. Right. Back to the drawing board.

At my last lunch at Dominique and Christiane's house, Christiane told me she loved the idea and had me write down a list of things she wanted me to achieve when I got back home.

"Trouve un ami français," she said. Find a French friend to teach me how to speak French better, so that the next time I came to France, she and I could talk to each other more easily.

Then, *"Trouve une ferme ou une boucherie pour un stage."* Find a farm or a butcher shop to study with.

"Fais un plan pour ouvrir ta boucherie." Make a business plan for my butcher shop.

And finally, *"Trouve un garçon riche,"* she said, laughing. Find a rich man to fund my butcher shop. "It will be much easier that way," Christiane said, only half-joking, I guessed. Easier, she meant, than navigating the narrow path that she and Dominique and his brothers had forged across that indeterminate expanse of known and unknown risks.

But I was no longer interested in *easy.* Kate and the Chapolards had helped me to see the potential rewards in their various guerrilla strategies, their own defiant form of ownership, and I had subsequently fallen in love with the hard work such a strategy required.

ON THE PLANE RIDE back to the States, my seatmate, a woman with bright-pink, angular cheeks, asked me where I was headed, employing the pitch-perfect, staccato English of a well-studied non–native speaker.

"Home," I said. "Oregon."

"And what do you do at home?" she asked.

I wasn't sure how to answer. I was far enough away from the world of magazine writing that I couldn't call myself a writer anymore. But I was hardly a butcher. I knew I was going home to unemployment checks, an embarrassingly low bank account balance, and a maxed-out credit card with a very high interest rate. And to the half-finished house I'd hastily moved into with Will. I was going home to all the reminders that my life had not turned out the way it was supposed to. I was going home to the person I once was. I had no idea who I was about to become. The space in between felt vast and lonely. I so desperately wanted to erase everything that had come before France and start over, like an immigrant in a completely new country. I even wished I could change my name.

And yet, beneath this pressing anxiety, I felt a near-innocent hopefulness, something I hadn't felt since I first moved to New York to become a magazine editor. France had snuck a tiny seedling inside of me. Its precise nature was still hidden. I didn't know what it needed in the way of nourishment, but I could feel it blowing around behind my rib cage, searching for a place to take root.

My seatmate waited patiently for a response, but I shrugged and turned away from her without answering and pressed my forehead against the tiny airplane window. As the plane took off, I looked out over the dirty-ocher expanse of Toulouse and silently repeated the vocabulary of French butchery I'd just learned. *Épaule. Saucisson. Filet mignon.* So many beautiful vowels with their soft edges and open-ended tones. I closed my eyes and imagined my life back home as one long, tiresome sentence made up entirely of consonants. I fell asleep briefly, jerking awake every few minutes, and in my spasmodic half sleep I dreamed that I spoke entirely in AEIOU.

This much was obvious: Learning to kill an animal and turn it into dinner had changed everything. So had picking up a knife and learning how to follow the road map of muscle and sinew and skin and bone. As did the realization that when you rub salt into meat and fat and let it sit, magical things happen, like bacon and *saucisson* and *jambon*. Except they weren't really magical. They were part of the real world of practical things, loyal citizens of the rapidly decaying empire of the genuine article. France had begun to feel like *real life* to me, and home felt like a supremely disappointing representation of it.

This was what I wanted my real life back home to start looking like: *We killed this pig. She was a good pig. I was there to help bleed her. He helped raise her. The pig lived a good life, had a good death. We will embrace the complexities of this experience and we will eat well tonight. We will savor every part and make it last. We will render the fat into lard and make salve and soap out of it. We will take those bones and heal ourselves with its broth. We'll dream in the language of fat and skin. This language will make us think. Thinking will fill us with reverence. This reverence will make us whole.*

John Berger once described the act of writing as resembling "that of a shuttle on a loom: repeatedly it approaches and withdraws, closes in and takes its distance." But unlike a shuttle, Berger goes on to write, "it is not fixed to a static frame. As the movement of writing repeats itself, its intimacy with the experience increases. Finally, if one is fortunate, meaning is the fruit of this intimacy." Writing had once, long ago, done that for me, and now learning how to turn a pig into dinner in France had forced me out of my own safe, static frame and into the realm of intimacy and meaning. That was the *real life* I wanted to keep living. But this is what I was returning to: A lone, pale pork chop in a Styrofoam package. Men in business suits feasting on bland twenty-nine-dollar filet mignon. Butcher shops with no one inside them who has ever even seen a live pig.

IN FRANCE I'd begun to develop a new lexicon for myself, one that didn't trade in fear or denial, one that embraced fully open-ended vowels, one that allowed for a more complex rendering of the world of food, one that would, inevitably, reach into other facets of my life as well. *Ands*, not *buts*. *Ands* that felt defiant.

Confront one genuine article and you're bound to encounter a whole lot more. My journey into the world of meat—a journey that forced me to dwell within that gap between what I thought I knew and what I had chosen not to, between who I'd thought I was and who I was becoming—inspired me to poke at every person, place, and thing around me until I revealed their inner workings. This maybe explains why my reentry back home looked a lot like a bad head-on collision.

PART 3

TWENTY-TWO

With its dust and gravel and potholes, the long dirt road that wound its way to Pure Pork farm, just outside Sandy, Oregon, was not unlike the long dirt road that led to the Chapolards' farm, except that where fields of féverole and barley bordered the Chapolards' driveway, fir trees lined Pure Pork's.

When I arrived, Levi Cole, a tall, lanky rope of a man, a few years younger than me, with a shaved head, angular cheekbones, and the kind of piercing eyes that slice through just about every form of bullshit, kissed me on one of my cheeks and then hugged me, saying, "Hey, lady! You ready to kill a pig?"

Levi looked down at himself. He'd torn the right leg of his jeans from just above the knee to just below his crotch, such that I could see his boxers, which were bright pink, with red hearts on them.

"What happened to you?" I asked.

"I got in a little bit of a tangle with that there rooster," Levi said, pointing to a rusty red rooster strutting around in the back of his truck.

"Do you happen to have an extra pair of pants, Linda?" Levi asked a sturdy woman in dirty Carhartt overalls, standing across the driveway from us. Linda Burkett stood quite a few inches taller than my five feet ten inches, sported short, dyed auburn hair, and was perhaps fifteen years my senior.

"She's the farmer," Levi told me. "Linda, this is Camas. Camas, meet Linda."

We shook hands. She had a confident, firm grip.

"Glad to meet you. So you're the reporter lady?"

"I guess so," I said, wishing that I were not, in fact, "the reporter lady," but just a regular person who'd shown up to help Levi slaughter the pig she'd raised for him.

"I'm gonna turn that cock into coq au vin next week," Levi said to me, pointing to his feathered opponent. "You gonna come over for dinner?" It would make a good scene for the story I was writing about him for a new local food magazine, so I said I would.

"I've got pants," Linda said to Levi. "They're ladies' pants, though. But heck, you'd look good in anything." She winked at me.

"I'll be right back," Levi said as they turned to walk toward Linda's house.

I'd met Levi a few years before I went to France, through Robert Reynolds, the chef and culinary teacher who'd introduced me to Kate and whom I'd profiled for the city magazine before I lost my job. For the story, Robert and I had cooked dinner together at his house for a few of his former students, one of whom was Levi. Before meeting Levi for the first time, Robert had described him to me as "the real thing, the genuine article."

LEVI — GENUINE ARTICLE, I'd written in my notebook.

Chatting with Levi over dinner that night, I began to understand what Robert had meant. Levi took everything several steps further than most people. He insisted on killing his own animals whenever possible, and he typically dealt with all the details of butchering them himself.

While Levi didn't own his own farm, he'd grown close with the farmers he bought pork, lamb, goat, and beef from, visiting their farms on a regular basis. In his urban backyard, Levi also raised his own bees for honey, chickens for eggs, and rabbits and chickens for meat. He hunted for elk and deer each fall. He fished. On a regular basis, he canned his own duck and rabbit rillettes, rendered his own lard, made his own soap. He processed his own pickles and preserves from the produce he grew in his garden. Levi was, for all intents and purposes, what many food-conscious people in Portland strove to be but were not entirely sure how to become: a self-possessed, self-taught, culinary obsessive-compulsive–cum–urban homesteader with a stomach for killing his own dinner. He funded all of this through his day job as a critical care nurse at a local hospital. He'd even refinanced his house so that he could study cooking in France with Robert.

Over a dessert of Robert's delicate crêpes suzette, Levi told me that every year for his birthday, he found a farmer to buy a pig from, killed the pig himself, and then roasted the pig for fifty of his closest friends. He also usually killed a second pig and butchered it himself so that he could make his own hams and bacon.

"I'll probably want to write a story about you someday," I'd told him. Two months back from France, I was doing just that.

While I waited for Levi to return, I roamed the farm and took notes. The ground was dry, save for the muddy pasture where Linda's pigs wallowed. I leaned on a wooden fence that separated all the pink pigs from me and watched one particularly large pig—almost as large as that seven-hundred-pound sow in the French abattoir—root around in the mud with what looked as close to glee as I could possibly imagine. I considered getting on my hands and knees to join her.

And I considered joining her because rolling around in pig shit and

Oregon mud sounded a lot more appealing than taking down notes while Levi attended to the real-world business of killing a pig. But I needed to write this story. I needed the money *now*. And I needed the money now because I'd been homeless, sleeping on friends' couches, and living out of my car, for two months, and though I'd finally saved up enough unemployment checks and scraped together enough writing gigs to be able to afford first and last month's rent on the tiny little detached garage studio I'd just moved into, I wasn't sure I could cover next month's rent.

Five days. That was how long I'd lasted back at Will's house after I returned from France. And then I'd left him in the middle of the night on the hottest day of summer, after waking him up from a deep sleep to tell him I couldn't do it anymore. *I am tentative about love and I am in love. I am afraid of being alone and I want to be alone.* This is what I'd wanted to tell Will when I woke him. Also, *I don't know exactly who I am anymore, but it's not who I was when I left for France.* But it didn't come out right. It was too complicated. *I'm done* would have to do the trick. Sever all communication between the heart and brain. It would be cleaner that way. Less pain, less suffering.

But nothing about that night was clean or painless. I ended our screaming fight by hastily packing a bag with a can of Jehanne's duck rillettes, a bottle of honeysuckle eau-de-vie, a toothbrush, and some underwear and slamming the front door, leaving Will, in his boxers, hunched over a creaky wooden chair in the middle of the living room to survey the aftermath of the hellish tornado that had just swept through his house.

"I want you to be happy," he'd said before I left for France. "Just don't come back from France and leave me."

"I'd never do that," I'd said. "I'd be a terrible person if I did that."

A LOCAL CHEF, Guy Weigold, arrived. We'd met only once before, at one of Levi's pig roasts. He was the owner of the Farm Cafe, a now shuttered, mostly vegetarian restaurant in Portland, and when he'd found out I was a food writer for the local city magazine, he'd joked, "Don't tell anyone I'm here. My customers probably wouldn't like it."

Three of Robert Reynolds's students, Porter, Tagg, and Nick, came around the corner of the barn. They were in their early twenties, dressed in that purposefully disheveled, hipster aesthetic—horn-rimmed glasses, tight jeans, plaid surf shirts, hoodies, Danner boots, Converse high-tops. They were earnest, respectful, curious guys who'd come on their own time, at Robert's urging, to witness Levi's pig kill.

A big white pickup pulled up just as Levi returned in a pair of Linda's dungarees, and a diminutive, dark-skinned man opened the driver-side door.

"Chief Dave," Linda exclaimed. "Hello!"

"Hello," Chief Dave said, not smiling. He looked around at all of us warily. "You got a lot of people here," he said to Linda.

Dave Strickland—his customers called him Chief Dave, although I never found out why—owned what Linda called a "state-inspected mobile slaughterhouse" and spent his days traveling from farm to farm to do what most people have little to no interest in doing: shooting four-legged farm animals in the head, bleeding them, then skinning and eviscerating them so that they could be turned into meat for the dinner table. Linda told me that smaller farmers like her, who weren't interested in running their own slaughter operation, preferred to hire an outfit like Chief Dave's to come to the farm to do the slaughter, as opposed to having to drive their animals to a large, unfamiliar, USDA-inspected slaughterhouse, which was often more than a hundred miles away for many farmers she knew. Linda also felt that on-farm slaughter was more humane, and more sustainable for her business, even if it legally limited her options for selling her animals. Due to various regulations I didn't yet fully understand, she was allowed to sell animals killed on the farm only in whole, half, or quarter form, directly to consumers who wished to use the meat for personal consumption. She was not allowed to sell to retail outlets like restaurants and butcher shops that, in turn, would be selling the meat for profit. To do that, she was required to take her animals to one of the USDA facilities. If you couldn't cooperatively own your own small slaughterhouse like the Chapolards did, this seemed like as good a workaround as any for a farmer like Linda. But it also meant she needed to find consumers who wanted to buy a whole, half, or quarter pig, who were as willing to take on the trotters and pig ears as they were the pork chops and bacon.

"Why spend a lot of money on gas and stress your animals out when you could have a guy with a gun show up?" she said. "The animals don't even know what hit them. They're happy one minute, dead the next."

Linda had hired Chief Dave to come and kill a pig for one of her other customers that day, so while Levi still planned on killing his own pig, he'd decided to pay Dave to show him how he handled the post-kill processing.

Chief Dave climbed out of his battered pickup, reached back into the truck's cab to grab a shotgun, and then turned to survey the barnyard. Having an audience was probably not customary for Chief Dave. Most people hired him to kill their animals so that they wouldn't have to witness any of it. We were all clearly city kids, tourists on his outdoor kill floor.

Levi approached Dave and shook his hand.

"Thanks for coming," Levi said. "So I've been killing a pig every year for about four years. The first year I tried to kill a pig, it was horrible. My friend Robert told me that in France they just hang a live pig up by its back feet and it falls asleep and they stab it in the neck. Robert is a good storyteller, for sure, but, as it turns out, live pigs don't actually fall asleep if you try and hang them by their feet."

We all chuckled, but underneath, I sensed a collective nervousness about what we were about to witness. Would it be much smoother than that?

After a couple more years of trial and error, Levi had perfected his method. "The whole thing is peaceful now, at least for the pig," he said. We followed Levi into the barn, where his pig had been settling in for several hours now, sniffing the walls of her hay-lined stall. Levi knelt down and massaged her cheeks.

"It's hard for us to watch. But now I feed them a half rack of beer, I make sure they are good and relaxed, I shoot them in the head, I bleed them out, and it's a happy ending." As he said this, Levi poured a bottle of Blue Moon Winter Abbey Ale into the mouth of the pig he was about to kill. It seemed as pleasant a way to go as I could imagine.

According to meat scientists I have talked to since, the jury's still out on feeding an animal alcohol before slaughter, but ensuring that an animal is rested and relaxed leading up to and during slaughter is in fact an important determining factor in the quality of the meat we eat. If, for instance, a pig is immediately slaughtered after stressful transport, the meat of that pig may be pale, soft, and watery—the result of fast rigor onset, which produces high lactic acid in the muscle postmortem, before the carcass can be chilled. The industry label for this kind of pork is PSE, or pale, soft, and exudative. Allow the pig to rest between transport and slaughter, handle the animal gently during slaughter, and chill the carcass rapidly postmortem, and PSE pork is much less likely. The opposite end of the spectrum is DFD or dark, firm, and dry meat, often due to extended activity or struggle or an inappropriate period of fasting prior to slaughter, which results in a depletion of glycogen storage in the muscle, leading to the muscle's inability to produce sufficient amounts of lactic acid postmortem. This meat may be too dry and is prone to spoil quicker. DFD is more common in beef, but can sometimes be found in pork.

"He sure loves his pigs," Linda said, nodding toward Levi, who had begun talking to his presumably tipsy two-hundred-and-fifty-pound Blue Butt, the common breed name for Linda's Yorkshire-Hampshire mix.

"Of course I do," Levi said.

WHISPERING SWEET NOTHINGS into a pig's ear was all good and fine, but killing a pig meant killing a pig, so Levi made a plan with Chief Dave.

"So I'm going to do the honors," Levi said, pulling a pistol out of his pocket. "But I want to see how you skin it and gut it."

"Fine by me," Dave said.

We grew silent. Levi let the pig smell the gun, so that this new, shiny object in her line of sight didn't scare her, and then he slowly moved the pistol up toward her head. A few seconds later, he shot her right between the ears. She immediately fell over sideways and began to convulse just like the seven-hundred-pound sow in the French abattoir had after she'd been stunned with electricity. Levi and Guy held the pig's body down so that she didn't thrash into the walls of the barn.

"When you shoot a pig in the head," Levi told us calmly, "its nervous system sends out multiple signals all at once, which makes the pig's body convulse. It looks terrible, but if I did it right, she can't feel anything right now." After she'd stopped moving completely, Levi and Guy lifted the pig's back legs up off the ground so that when Chief Dave stuck a knife into the carotid arteries, the blood ran toward the ground. No one was there to catch it.

The sound of blood is more shocking than the sight of it. I could hear it spilling out onto the sweet alfalfa beneath the pig's body. In *Pig Earth*, Berger describes a cow being slaughtered. "Life is liquid," he writes after seeing its throat cut. "The Chinese were wrong to believe that the essential was breath."

Levi and Guy set the pig back down on the hay-covered ground. Levi turned to me. I could see the pulse in his neck tracking time. "Every year I do this, it's a little horrific. The day it isn't, I don't think I should eat pigs anymore. Write that down in your little notebook, Ms. Reporter Lady."

In a mere five minutes, Chief Dave hung the pig by its back legs, skinned it, and then sliced down its belly and eviscerated it, letting the pig's innards slide out into a large bucket.

Levi explained to Chief Dave that normally he and his friends plunged the pig into a vat of scalding hot water for a minute and then shaved the hair off with knives so that they could save the skin, too.

Chief Dave shook his head. "Seems like a whole lot of work for nothing."

"But the skin is delicious," Levi said. Levi, like me, had been to France, had learned to use the skin to thicken his soups and stews.

"Make sure to get the leaf fat," Linda said to Levi.

"What's leaf fat?" one of Robert's students, Nick, asked.

Leaf fat, Levi explained, lines the abdominal cavity and the kidneys, and is prized because of its high lipid count and creamy consistency. It also burns at a lower temperature than fatback, which is where traditional lard comes from, he told us, and so, in its rendered form, leaf fat is perfect for pastries and pie dough.

"It's what people used before Crisco," Levi said.

The Chief gave Levi a suspicious look. "What are you going to do with it?"

"I'm gonna make pies all year long, daddio," Levi said.

"You're keeping the head, too?" Dave asked.

"Is that uncommon, Dave?"

"For white people, yeah."

"I'm not your average white guy," Levi said, and pointed to the bucket of guts. "I'll take that heart and liver, too."

Chief Dave went over to the bucket of guts on the ground, cut the heart away from the rest of the organs, and handed it to Levi. Levi offered it to me.

I tucked my pen behind my ear, shoved my notebook in my back pocket, and stood there with the pig's heart in my hands.

TWENTY-THREE

Two days later, after Levi had hung the pig in a friend's restaurant walk-in and the carcass had gone through rigor mortis, our same group—Guy, Tagg, Porter, Nick, and I—gathered around a stainless-steel worktable at the studio where Robert taught his students.

Levi had a busy few days ahead of him. His to-do list for the weekend read something like this: (1) butcher pig; (2) rabbit rillettes; (3) duck confit; (4) kill roosters; (5) render fat; (6) make sausage; (7) salt pancetta and guanciale; (8) start posole; (9) cut soap; (10) order beef from Larry; (11) feed bees.

He began walking the students through what they would do with the pig liver and head.

"We'll make pâté today out of the liver. I'll make guanciale out of the jowls. Jowl bacon, in other words. And I'll crack open this head, take the brains out, and put the entire thing in a stockpot and make posole," Levi said.

Levi washed the liver under cold water and then gathered the students around him. He cleaned the liver using the same knowledge he relied on every day as a nurse, offering words like *portal*, *lipase*, *vena cava*, and *common bile duct* to explain what he was doing.

"This isn't rocket science," he kept telling the students. "It's a body, just like ours. This is a liver, just like ours."

Not too long ago, Levi said, this was a common annual ritual among communities all over the world: slaughter the hog, butcher it, and then preserve the meat by making hams, bacon, sausage, and pâtés.

"Ask Camas. She just got back from Gascony, where they *still* do that," Levi said.

For this group of urbanites, however, it was new territory. For us, a book on slaughter and butchery—which Levi had placed on the worktable in front of us—stood in for a grandmother in an apron and a grandfather with a shotgun.

After preparing liver pâté and finishing off the head, we broke for a lunch of meat, cheese, and bread. Robert arrived and poured us some wine, which prompted Levi to tell us about his childhood.

In Estacada, Oregon, on the communal farm—"a pot farm," he told us, "that eventually got busted"—where he grew up in the seventies and eighties, with nine adults and five kids, they raised and killed their own animals for food. "The first time I was part of killing an animal, my uncle told me to plug my ears. I was six. He pulled out a pistol, shot a goat in the head, and then hung the thing up and asked me to hold it still for him. I gave it a big hug while he tied its feet."

Levi's friend John Taboada, a local chef who did whole-animal butchery occasionally in his restaurant, Navarre, showed up. He'd agreed to show us his butchery method on one half while Levi followed along with a knife on the other half. We all joined them around the table. From a folded dish towel, John unwrapped what looked like a small paring knife and began cutting the tenderloin away from the vertebrae and ribs, just as Dominique had shown me.

"I read once that when the lion kills its prey, the first thing it eats is the tenderloin," John said. "The tenderloin is what I always take out first, because it's valuable, and you could accidentally cut it as you start working on other parts." Lions. French grandmothers. Rich American businessmen. The story of the tenderloin was getting more and more complicated.

John's knife scraped against bone. "Hear that? That's what you want to hear when you're butchering."

"In the United States, there are something like twelve basic cuts on a pig," Levi said. "In France, there are something like forty-eight. Right, Camas?"

"Something like that," I said, although I didn't actually know and it seemed to me that the number depended on the butcher and what he or she was going to do with the meat. Perhaps there really was no such thing as American cuts and French cuts. There could only ever be Dominique's cuts or Levi's cuts or John's cuts or my cuts—cuts that made the most sense for the people who would eventually eat them.

"My friend in Italy keeps trying to get flank steak from his butchers after I cooked it for him in my restaurant," John said. "They have no idea what he's talking about. It's different everywhere you go."

"I go to Haiti every year to do volunteer medical work," Levi said. "They just hack the carcass into rough pieces and stew it all. There's no time for anything else. So many people don't have refrigeration, and everything needs to be cooked immediately."

Next, John cut the front and back legs away from the middle portion. He did not use a cleaver or a saw, only that small knife inserted into the spaces between joints.

"No one trained me," John told us. "I just figured it out myself." Levi had mostly taught himself, too, learning from others when he could, and from books, but mostly it had been—and to some extent it still was—a process of trial and error, with a few mentors like John and Robert and Chief Dave to guide him in between.

While they cut, Levi asked the students what all of this had meant to them. Porter put it best. "I was in architecture as an undergrad, and I understood, by looking at pictures and plans, what a building looked like. But there's an experience of going into the building that's completely different. That's how I feel about this pig."

"What about you, Ms. Reporter Lady?" he asked me.

I paused for a minute but did not answer. I had so much to share about what I had learned in France, but I was here to report a story, not to tell my own. I picked up a spare knife, and set about seaming out the ham.

BY THE END of the day, we'd cut and vacuum-packed everything but the belly and the meat that Levi would turn into sausage. Levi would take care of these parts later, because there were still rabbits back home that he needed to process, and bees to check on, and he still needed to kill that rooster he'd bought from Linda, plus some other roosters he accidentally had in his backyard—some chicks he'd bought turned out to be male—but wasn't supposed to, since roosters were illegal in Portland due to their prodigious vocal talents. I followed Levi to his house to help him.

Levi's North Portland backyard was muddy and full of sweet-smelling rabbit manure and firewood and food: eggs, chickens, lettuce, squash, carrots, chard, rabbits. Levi boiled some water in a large pot on a propane burner and then grabbed a rooster and instructed me to hold it upside down by its feet to calm it. He said this curtly, as if he were working on a patient at the hospital and didn't have time to joke around. I sensed he was done talking to Ms. Reporter Lady for the day, so I put my notebook and pen away, grabbed the rooster's feet with one hand, and wrapped my other hand around its wings, as Levi instructed me. Levi pulled a small knife from his back pocket and stuck it into the top of the rooster's mouth, just as they had at Jehanne's farm.

"That was to scramble his brain so he can't feel what comes next," Levi said.

He then stuck the knife into the rooster's neck.

"That was the carotid artery I severed," he said. "Now we wait for his nervous system to completely shut down." As the rooster bled out, it shuddered a few times in my hands, its claws drawing circles in the air as if it were running in slow motion, much in the way our old hunting dog, Gabe, used to move his feet while sleeping, dreaming of flocks of geese, most likely. Levi touched one of the rooster's eyeballs with his finger.

"If the rooster flinches when I do that, it means he can still feel pain and I did it wrong. But he didn't flinch. That blinking is just part of his nervous system making its last stand."

My heart pumped a little faster as I watched the rooster's eyes close for the last time.

Four days earlier at Linda's farm, I'd noticed a rooster in the barn, standing on one leg, one eye closed, one eye open.

"Wild ducks will line up in a pond on a log," Levi had told me. "All the ducks in the middle will be asleep, and the two ducks on the end have their outside eyes open, their inside eyes closed. They're keeping watch. Their brains are half-asleep, half-awake.

"Chickens sometimes do something similar when they sleep," Levi said. "They stand on one leg and keep the opposite eye open."

I imagined all the meat eaters in America lined up like ducks in a pond on a log. We're all huddled in the middle, sleeping. Maybe Linda or Chief Dave takes up one end. Maybe Levi takes up the other. And as we sleep, they keep watch. They see. They know. They carry the weight of a certain reality for us so that we don't have to.

Levi killed a few more roosters and then sprayed their carcasses with a hose. We dipped them in hot water. We plucked their feathers, removed their heads and feet and guts. The next day, after they'd gone through rigor mortis, Levi would tuck one into a pot of simmering wine with onions and carrots, garlic and peppercorns.

"Here's to dinner." Levi held up a glass of beer.

I took a long swig of mine and sat down on one of the wooden steps of his back porch.

Levi sat next to me. "So what *do* you think about all this?" he asked.

I was still on unemployment. I maybe couldn't make next month's rent. I was now doubly heartbroken, living alone, and unsure of what the future held for me. I was back to writing again, but I didn't want to be. France seemed like a faraway dream. Christiane's words still haunted me: "If you don't do anything with this, we will be very sad. It will all be for nothing."

I thought about what Robert's student Porter had said, how the experience, as an architecture student, of going into a building was totally different from looking at pictures and plans of a building. Porter had said he wasn't quite sure yet how watching Levi kill a pig had changed him. He just knew it had.

"I have an idea," I said. "Maybe you can help me."

TWENTY-FOUR

Surely, I explained to Levi, we weren't the only people in Portland searching for this sort of transparent experience, who wanted to take part in the process of getting meat to our tables the right way, the old way, the respectful way—whatever you wanted to call it. The real way, if such a reality was even possible anymore. Surely we weren't the only people unconvinced that the answer to the entire meat conundrum was as simple as forsaking meat altogether. Surely we weren't the only people who would rather eat meat from farmers like the Chapolards or Linda than from anonymous factory farms four states over, or four countries over, for that matter.

All around me in Portland, I saw people hungry for connection with their food. I saw words and phrases like *sustainable*, *local*, *organic*, *farm to table*, and *humane* tossed around at farmers' markets and grocery stores, on restaurant menus. I myself had used these terms in my food writing on occasion. But what did those labels mean, and did they really connect us to the thing itself—the act of killing and butchering an animal for dinner? If more of us were connected to the thing itself, how might each of us become a very different kind of meat eater?

Even outside of the politics of food, the economics of the plate, the lure of sustainability's lingo, I knew there was something deeper that would resonate with people in Portland, just as it had for me in France. It was learning how to do something tangible, meaningful, and direct with my own two hands. If anyone asked me where the ham on my plate came from, I wanted to be able to tell them, from beginning to end, how it had gotten there. *That* was a kind of ownership—maybe not the kind of total ownership the Chapolards had achieved, but certainly a reclamation of knowledge and skill that had been taken away from us once industrialization took over our food system.

I didn't want to disappoint the Chapolards, but I was far from ready to open a butcher shop or start a farm, if not solely because I had no money to do so. Plus, I wasn't even convinced that the Chapolards' model would work back home. I needed to keep learning. I wanted to make a living, somehow, outside of writing. I wanted to re-create my French education in Oregon, not only for myself but for my community.

What if, I thought, by bringing people closer to the basic, old-fashioned processes by which meat got to our tables—by drawing the comparison for them between that 99 percent of animals raised on factory farms and that other 1 percent—I not only inspired them to change their buying and eating habits, but also to change the way they saw the entire world, their community, and everything else that stood between life, death, and dinner? What if it made people grapple in ways we modern, urban beings were rarely, if ever, forced to grapple? Maybe simply being told how awful industrialized farming could be wasn't enough to make people change how they ate. Sure, the industrialization of our food system had, in many ways, severed our brains from our bodies, allowed us not to think about what we ate, but what if some people wanted things to be a little more difficult—what if more people wanted to think and register and feel and see what happened?

In *Eating Animals,* Jonathan Safran Foer writes, "There is something about eating animals that tends to polarize: never eat them or never sincerely question eating them; become an activist or disdain activists."

I'd been on both sides of this equation. But now I found myself in the middle of the gap between death and dinner. I was on the search for *and* people, not *but* people. Levi was one of those people. So were Linda the farmer and John Taboada. I needed to find the Chapolards of America. I wanted more teachers. More mentors. More allies. More people hungry, like me, to learn and open our eyes and see.

It was a rather lofty ideal, of course, but, sitting there on Levi's back steps, I began to pull together a business idea: a fully transparent, hands-on meat-education-and-buying model I eventually dubbed the Portland Meat Collective. I'd source animals from local farmers like Linda. I'd use those animals in classes that encouraged people to pick up knives and learn how to slaughter, butcher, cure, and cook every part of those animals for food. I'd hire people like John and Levi and Chief Dave to teach the classes. And although I wasn't quite ready to teach myself yet, I'd at least be able to help those teachers interpret for our students what they were doing with their knives, and why. Maybe the classes would even prepare me to someday be able to teach others myself. Maybe, eventually, the Chapolards and Kate could come and teach. We'd show students how to use every part and make it last. We'd make *pâté de tête* together and maybe no one would cringe. And then the students would go home with all that meat to fill their freezers and make their own *jambon*. Just as Levi had said, people used to do this all the time with their elders, their neighbors, their families. I was simply going to re-create this scene for modern, urban times, give it a name, register as a business with the state, maybe even pay myself for the work it took to bring this sort of experience to others.

I told Levi about my idea and asked him if he might consider teaching a class.

"I'm in," he said.

OF COURSE, I had absolutely no money to start a business, so I set about scrapping. I convinced a designer friend from my magazine days to create a brand identity for me in trade for meat and classes down the road. My brother, who ran a successful Web design company, generously helped me throw together a Web site for free. A friend recently out of law school acted as my pro bono lawyer. People liked the idea and were happy to help.

I called up the United States Department of Agriculture and the Oregon Department of Agriculture and every other relevant Department of Fill-in-the-Blank that I could find and asked whoever was willing to listen whether what I wanted to do was legal.

Sort of, they said. *Maybe? It's a gray area. No one's ever asked us this before. There is simply no precedent.*

No precedent for a farmer, a butcher, and a bunch of regular old people gathering together around a table to turn an animal into dinner. At least no precedent that anyone could remember.

Was there some license they could give me to make what I was going to do legitimate?

We don't have a license for what you're about to do, they said. Hearing this thrilled me. What I was about to do felt defiant in all the right ways.

So could I do it?

So long as you walk very carefully between about one hundred regulatory lines, they said.

I could do that. But first I asked them to tell me everything they could about the nature of these lines they were talking about. And then I put my old magazine fact-checking hat on and set about diving deep into the gray area, feeling out its boundaries, asking a whole lot of questions.

In just over a month, word traveled quickly. Whether it was because I knew the right people, or more people had been watching me from afar than I cared to know about, before I'd even registered my business name with the state of Oregon, before I even knew how I was going to pull off the first class, the local media started calling. They asked me questions like "What does it mean to be an 'ethical' butcher?" and "How transparent will your meat classes actually be?" I stumbled over the answers—my new lexicon was still in progress. I didn't even consider myself a butcher yet. I wasn't even sure how to go about making my idea a reality.

But I kept meeting the Chapolards and Kate Hills of America. They kept offering me help.

And then I fell in love with one.

TWENTY-FIVE

It was the end of the day, already dark outside, and when I opened the door to Pastaworks, a European-style market on Hawthorne Boulevard in Portland, a gust of cool October air ushered me in.

As a teenager, I occasionally drove up to Portland from Eugene for the day with friends who had their own cars. We'd strut up and down Hawthorne in our vintage velvet dresses and batiked leggings, poking our heads into tie-dye and incense stores with brightly colored Indian tapestries in their windows. Today Hawthorne is home to an American Apparel and a Ben & Jerry's, but the scent of Nag Champa still wafts from a few of those same stores, their window mandalas long ago faded by the sun. Every time we walked by Pastaworks, I'd press my face up to one of its windows and peer in at all those exotic-looking ingredients. It seemed like a place meant for adults, adults with money, adults with an understanding of the world I had no ability to comprehend.

After arriving home from France, however, I'd discovered that, aside from Portland's farmers' markets, Pastaworks was as close as I could possibly get to the markets I'd fallen for in Gascony. Not *quite* the same—it was, in truth, more Italian than French in terms of its ingredients, it was more American than Italian in terms of its prices, and no one wagged their finger at me when I picked up a melon and smelled it—but it would have to do.

Upon entering Pastaworks, you were immediately hit with the funk of olive brine and aged cheese. On the left side of the store, a wide, deep refrigerated trough spilled over with aged pecorinos and nutty blues, tart and crumbly chèvres, sharp cheddars, luscious rounds of raw cow's-milk cheese wrapped in thin strips of spruce bark, and thick, oozing triple creams that, back when I could afford them, I ate with a spoon for breakfast. Off to the side, vats of salty olives seasoned with rosemary, jars of tiny pickled anchovies, called *boquerones*, with glistening silver skin and creamy flesh, bowls of bright roasted red piquillo peppers preserved in olive oil. In the middle of the store, you'd find a copious bounty of bright-green broccolini, striated purple-and-white radicchio, or, depending on the season, velvety mâche, all grown by farmers only an hour or two away. On the other side of the produce mountain you might discover wild fiddlehead ferns waiting to be sautéed in sweet butter and green garlic or, in the right season, a pile of plump matsutakes plucked straight from the wet, pungent soil of Oregon's Willamette Valley by a crazy-eyed mushroom forager named Lars. At the back, owner Peter de Garmo, a white-haired, bespectacled, quiet man who opened Pastaworks in 1983 and probably still knows more about food and wine than all the denizens of Portland combined, sat at a tall desk, surrounded by shelves of Italian Barolos and Barbarescos, bottles of French Chinon and Bordeaux, obscure Oregon pinot noir and pinot gris. And on the right-hand side of the store, a vast glass butcher's case stacked with thick T-bones and loin chops, beautifully tied pot roasts, breaded scaloppini, and plenty of tenderloins.

I was surprised to see a woman standing behind the meat counter, leaning against it with her right elbow, her chin resting in her hand, talking to her male co-worker, a rockabilly type with meticulously groomed sideburns. She spoke with a honeyed twang. She wore a baseball cap pulled low over her dark-brown eyes.

Before I went to France, this had been one of the meat counters that rejected me when I asked if they'd take me under their wing. Only men had been working behind the counter then. The woman stopped talking as I made my way toward them. They wore white, button-up butcher's smocks, and hers was two sizes too big on her petite frame. She punched him on the shoulder and, under her breath, said to him, "I got this one."

The meat counter was situated a step or two up from the floor I was standing on, such that the meat in the case could be viewed at eye level, and anyone who worked behind the counter towered at least a foot above my head. Because of this, even though she was clearly slight in body, she cut an imposing figure.

She looked my age, maybe a little bit older, judging by the crow's-feet around her eyes. Her jaw was square, her face long. I caught her eyes briefly, and in response she tipped her chin up and out, pursed her lips, and stared at me from under her baseball cap, almost as if she were challenging me to a fight. I lowered my gaze back to the rib eyes and tenderloin roasts between us. There were no pig ears or trotters here. No blood sausage.

"Huhney. What can I do you for?" Unlike her somewhat menacing presence and expression, her soft low voice and her Southern accent soothed in the kindest of ways.

"I'm looking for a pretty rib eye."

"Whaddya mean by pretty? Which one do you like?"

"Maybe one with a lot of nice fat. Maybe that one." I pointed to a bright-red rib eye with a beautiful white cap of fat around the edge and what I deemed to be proper marbling, though, in truth, at that point, I really didn't know much at all about beef or marbling. It was what I would call a classic rib eye. The kind you see on American grocery store billboards that say something like WE VE GOT USDA PRIME!

"Those are corn-finished. But we also have grass-fed. Not as much fat around the edges, but just as pretty, and if you ask me, they taste better," she said.

I'd heard the term *grass-fed* tossed around back when I was writing about Portland restaurants. I knew it was more expensive than grain-fed, but I didn't know why. I knew it meant that the cattle ate grass at some point in their lives, or maybe their entire lives. I knew it was a *thing*, a term people in Portland, Oregon, had glommed on to, as they had with *organic* and *natural* and *cage-free*. But I didn't really know *why* it was a thing, why it might taste better or *be* better. France had set me on a course to knowing, but I was a long ways from really knowing.

"I need it for a photo shoot, so it just has to look appealing."

"Huhney, do I look like the kind of person who would put an unappealing rib eye in my case?" She winked at me.

I surprised myself by letting out a flirtatious, demonstrative, engaging laugh. Since my return from France, I'd grown quiet and morose. I barely recognized that laugh as my own. It was as if someone much more at ease than me had taken over.

"What's the photo shoot for?"

"I'm starting a meat school, sort of." My designer friend had agreed to produce a promotional postcard for me in exchange for meat, and the plan was to shoot a rib eye on a crinkled piece of butcher paper with some butcher's twine.

"Get out of here! A meat school?"

"I just went to France to learn butchery, and I want to keep learning. And since there's nowhere to learn . . . Anyway. It'll be for anyone who wants to know where meat comes from. Totally hands-on. We'll use whole animals from local farms."

"Wait a minute, what's your name?"

"Camas."

"Camas what?"

"Camas Davis."

She sucked air in through her mouth, put her hands up on top of her head, and opened her eyes wide. "Are you fucking kidding me? You're that food writer who went to France!" She covered her mouth with her hands and looked around the store sheepishly. "Sorry," she said in a hushed voice. "I have a swearing problem. HOH-LEEE-SHIT. You're Camas Davis. *THE* Camas Davis? I've been reading your blog. I know all about you!"

A couple of other customers had gathered around the meat case and were listening in on our conversation. She held up a rib eye for me to look at. "What do you think about this one?"

The meat was a beautiful deep red, but there was hardly any fat on it, and the fat, in my estimation, was what made a rib eye look like a rib eye to Americans, even if most people never wanted to actually eat the fat. I did not yet know that, while feeding corn to cattle is what gives rib eyes their signature fat-to-meat ratio, cattle, being ruminants, aren't so adept at digesting corn. I had some sense that feeding cattle corn made them sick, and that this led to the overuse of antibiotics, and somewhere in all of this we had a methane-production problem on our hands. But I didn't know that the fat this corn diet resulted in often tasted generic. That we were basically feeding cattle something they weren't meant to eat in order to raise mild-flavored meat with a marbled look that some marketing concern had taught us to value, when ultimately we didn't even want to eat the fat because at some point some nutritionist had told us animal fat was bad, which, recently, has been debunked by those same nutritionists.

I also did not yet know that a truly grass-fed rib eye contained a different kind of fat—fat with a much healthier ratio of omega-6 to omega-3 fatty acids than grain-fed beef—which resulted in a more complex flavor, the complex flavor of, say, clover and ryegrass, foxtail and fescue. All I wanted at the time was a classic picture of a fat-capped, marbled rib eye that everyone, at least in America, would look at and want to eat.

"I think I want to go with corn-fed," I told her. It was cheaper, after all, and my pockets weren't exactly spilling over with dollar bills.

"All right. Suit yourself. But grass-fed is where it's at. Let's see. This one has a nice fat cap on it. Does that work?"

"Perfect."

She tore off a sheet of brown butcher paper, and when she'd finished wrapping the meat, she leaned over the counter as far as she could and motioned with her hand for me to get as close as possible. I stood on tiptoe and leaned in.

"You ever tasted *jamón ibérico* before?" she whispered. I liked how she attempted, and failed in the most charming of ways, to combine her Southern accent with a Spanish one to say *"jamón."*

"No," I whispered back. "But I was just in Spain and I got to taste a lot of other kinds of ham."

She looked up and scanned the store.

"Hang on a sec."

To the right of the fresh meat in the case, they'd stacked all manner of charcuterie. I recognized the oblong dried coppas, with beautiful stars of fat running through them, because they looked just like those the Chapolards sold, although the people behind the counter hadn't made any of this charcuterie. There were also pink-and-white-speckled Italian *finocchiona* and Calabrese salami, Spanish chorizo chubs, Italian prosciuttos, German *landjäger*. Even though I'd witnessed the effects of time and air and salt on meat in France, I still had so much to learn. What was the difference between a *finocchiona* and a Calabrese salami? Did they taste anything like a French *saucisson*? Why did Spanish *jamón* have the trotter still on, while Italian prosciutto didn't?

She pulled out a whole leg of Spanish *jamón ibérico*. The price tag that sat next to it said $99/LB. She placed it on a shiny metal slicer, flipped a switch, furrowed her brow in concentration as she sliced it, and then handed me a crinkled piece of wax paper with a thin sheet of salty ham on it.

"Shhh. Don't tell," she said.

I put the whole slice in my mouth.

"Formidable," I said and smiled.

"Formidable? What the hell does that mean? It's fucking amazing is what it is," she said, again in a higher octave and loud enough for the other customers to hear. "Shit. Sorry for swearing, ma'am," she said to the customer standing closest to me. "Anyway, I guess these pigs are finished on acorns and half-wild. I've been talking to a couple of farmers nearby who are trying to feed their pigs hazelnuts. I wanna learn how to make this someday."

"Me, too," I said. We stood in silence for a second while I chewed and swallowed and shook my head up and down and said "Mmm" and "Yes" and "This is delicious."

She cleared her throat, shifted from one foot to the other, and cracked her knuckles. "Anyway, umm, do you like whiskey? You look like you like whiskey. I wanna take you out for whiskey sometime and hear about this meat school of yours."

I blushed. Was she asking me out on a date? Was she flirting with me? I hadn't flirted with a girl since I attended college, at a small, mostly female liberal arts school in Ohio, where, at a party one night, over a six-pack of Labatt, a surly, loud lesbian named Jules informed me that I was known as the "most-wanted straight lesbian on campus." Plus, even if we were flirting, I was in no position to be flirting back. After what had happened with Tom and then Will, I didn't want to subject anyone else to my deep and growing ambivalence toward any form of romantic love.

But still, I felt giddy. Giddy to have met another woman who, like me, not only had a penchant for whiskey and a keen interest in *jamón ibérico* and rib eye fat caps, but a woman who, unlike me, had actually landed a job behind a meat counter.

"I love whiskey," I said, a little flustered. "Yeah. Sure. Let's get together. I'm going out of town for a bit, but give me your number."

She tore off a small piece of butcher paper and wrote her number on it.

"My name's Joelle. But you can call me Jo," she said, and winked. "Promise you'll call me?" she asked.

"I promise."

"Because gals like us, we gotta stick together. Plus, I wanna help you with your meat school."

Had I just met my doppelgänger, but with a Southern accent and a propensity to swear even more than I did, who wanted to figure out how to make her own hazelnut-finished ham?

She handed me the wrapped rib eye, and as I made my way to the cash register to pay with the last dregs of that month's unemployment check, I tracked her movements out of the corner of my eye. When I finally turned toward her again, she was staring back at me.

As I walked toward the front door, she yelled out, "See you soon, Camas Davis."

TWENTY-SIX

I met the other love of my life three days later. One week before I met Jo, some married friends of mine, Hava and Scott, had called up at the last minute and invited me to join them and their friend Andrew on vacation in Hawaii. My financial situation was still rather shabby. The unemployment office had admitted me into a new self-employment program that allowed me to collect an unemployment check while also making money as a freelancer, so I wasn't living in total poverty. But my income stream was still tentative at best. Plus, I was spending every extra dollar I had to pay off my credit card or pull together the Portland Meat Collective. Nonetheless, I managed to find one of those suspiciously cheap last-minute airline deals and, against my better judgment, pulled out that magic credit card again. I figured I'd pitch a few Hawaiian food stories in order to justify my trip, and if that didn't work out and I had to declare bankruptcy and go back to living on friends' couches, so be it.

I got on the plane, drank the complimentary syrupy mai tai that the flight attendant handed me, and watched Tom Hanks hug a volleyball on the tiny television screen. When we landed, Hava, Scott, and Andrew were waiting for me.

I lugged my suitcase up to the white Jeep they'd rented and hugged Hava and Scott. Andrew and I cordially shook hands, mutually remembering out loud that we had met each other before at various parties Hava and Scott had thrown over the years. The three of them had arrived the day before, and Andrew was already bright red from sunburn.

"Looks like you got some sun," I said.

"This is actually my natural color," he said, attempting to look offended, but then Hava chortled and I found myself joining her.

ON THE DRIVE from the airport, Andrew turned up the radio in the Jeep. The Bee Gees were on, and he unabashedly sang all of Barry Gibb's high notes, with perfect pitch. We drove up and down a few steep hills, past vast fields of black, hardened lava, and up to the house they'd rented, where we sat by a tiny, kidney-shaped pool, drinking gin martinis and catching up. They asked me about France, but not too much. They asked me about Will, but not too much. I needed that kind of distance. I was still fairly incapable of talking about anything that had happened in the past few months, and they instinctively understood this, so instead we made fun of Andrew's hair, which he'd grown out into a tangled half mullet. Hava and I watched as Scott and Andrew attempted to toss a ball around in the pool. Slowly, I began to *see* Andrew. The ever shifting green and brown colors in his eyes. His particular way of playing the fool for the delight of those around him. The way he laughed at himself. His hands, which seemed capable and sturdy. Hava and Scott jokingly called him Colgate, not only because his teeth were so straight and white, but because he had an all-American, clean-cut, Midwestern air about him that didn't quite fit into the counterculture fabric of Eugene, where they'd first met him. Yet he always buttoned his shirts wrong, such that one side was longer than the other, and he didn't really bother to fix it when someone mentioned it to him. I saw the way he'd stare off into the distance while someone else talked, as if completely uninterested, but would jump in at the perfect moment with the most appropriate, authoritative retort. His smile radiated ease and kindness, as well as a bit of mischief. He possessed an incessantly restless energy. I noticed, too, that he was smart and curious, though he tended to hide these qualities behind a veil of goofy Midwestern charm. I was also, almost immediately, wildly attracted to him.

As it turned out, Andrew, like me, was freshly single. But I was nervous to insert myself into anyone else's life at that point. Perhaps that was why, even after I began to *see* him, I avoided talking directly to him without Hava or Scott in the room. Andrew mirrored my behavior. If we were caught alone, one of us would dart over to another part of the room and busy ourselves with the television or a book.

The four of us spent the evenings trying to come up with a tagline for my new business. Andrew, who had an undergraduate degree in math and a master's in business—foreign languages to a girl with degrees in comparative literature and performance studies—kept asking me: "But how are you going to make a living doing this?" I didn't know how to answer, and I resented the question. The Portland Meat Collective wasn't about the money, I told him. It was about cultural defiance, about shifting our social attitudes toward food. Years later, I would come to appreciate it, his insistence that I assign monetary value to my time and ideas, our theoretical arguments about the importance of social versus financial capital.

Andrew seemed at ease in any situation, and his presence had a grounding effect on me. The more he grounded me, the more attractive he appeared to me, despite how different we were from each other.

Three days in, we were flirting underwater and sleeping in each other's beds.

ON OUR LAST DAY on that black, burned volcanic island, the four of us snorkeled off a beach near the airport, right before our flight. I dreaded going home. Even with my various writing assignments and my plotting and planning the Portland Meat Collective, I still had so much time on my hands, and I was unsure exactly how I should fill it. I also didn't know whether I would see Andrew again or if we'd just part ways. I tried not to care either way, but I was already imagining what something beyond a tropical tryst might look like with Andrew back home.

I hadn't snorkeled before, and once in the water I kept lifting my head out to see where I was in relation to the shore. When I finally managed to keep my head underwater for more than thirty seconds, I would swim away from whole schools of vibrantly colored fish in order to go find a better school of fish, only to find nothing.

Andrew swam over.

"You okay?"

"Yeah. I just feel like I'm not doing it right." As I said this it occurred to me that I was maybe the only person in the world for whom snorkeling felt stressful.

"Just let go and float there and see what swims in front of you."

I put my mask back on and swam a ways out, tracking Andrew's body in the water. He stopped moving and so did I. We floated there on top of the water together, looking down into the abyss below us, dotted with bright-orange and pink coral. Andrew grabbed my hand and pointed to the right of me. A sea turtle floated below me. I looked back at Andrew, but he was gone.

And so I did what Andrew told me to do. I let the waves move me, and the turtle floated right there with me as each wave gently rocked us farther along. Occasionally it moved one of its flippers to steer, but mostly the turtle seemed not to be going anywhere in particular. After some time, I felt as though my body and the turtle's were moving in sync, both of us carried along by the lapping of the ocean, and this served to shut my brain off completely.

The only other time I'd felt this way had been several years before, when Jill, who had been fired alongside me, convinced me to go skydiving with her for a story she was writing for the city magazine. As we went up in the plane, sweat poured from every part of my body, my hands shook, my mouth went dry, and my tongue stuck to the roof of my mouth. I was possessed, in every physical way possible, by fear.

"I don't think we should do this," I said to Jill, shaking my head vigorously. "I can't do this."

And then, without much warning, a young guy clipped himself to my back, yelled "Here we go!" and pushed us out the door of the plane. We were falling, and because my brain could do absolutely nothing to stop this rapid free fall toward the ground below, it simply ceased functioning. Either I was going to die or I was going to live. And I felt, maybe for the first time ever, total and utter relief. I was just a body falling through space.

When the guy attached to my back finally pulled open our parachute, I laughed in a way that sounded a tad deranged. I was crying, too.

"Thank you. Thank you. Thank you."

He said, "You're welcome," but I wasn't really talking to him. I was talking to my brain, which had released me momentarily from its incessant tyranny. I was high for a week afterward, with a voracious appetite for chocolate and black licorice and sex. I would lie on my couch, think nothing, and shudder at the monumental sensation of my own body just existing in the world.

Here in this blue water, floating above the turtle, which was letting itself be carried by water toward something I would never know, I felt an exhilarating, wild *moving toward something*, without having any control over what that something might be. My body, there, alive, real, existing, moving, breathing. The turtle's body, there, alive, real, existing, moving, breathing. Andrew's body somewhere nearby, alive, real, existing, moving, breathing.

TWENTY-SEVEN

Back at home, I had a lot to do to make the Portland Meat Collective a reality, like finding farmers to purchase whole and half animals from, and butchers and chefs to teach the classes. Plus, there was Livestock.

Soon after I returned from France, a local public relations woman I knew from my city magazine days had reached out to me.

"Are you back? I heard about your time in France. I've got this idea for an event series. Meat authors discuss the hard stuff at restaurants. I'm calling it Livestock. Let's get together."

Lisa and I met for lunch a few days later, and I told her about my idea for the Portland Meat Collective—in fact, she helped me crystallize the name and the mission. She shared her idea for the author series. That all sounded good and fine, I said, but why not add a demonstration of whole-animal butchery to the mix? And invite farmers, too?

The plan we came to was this: In front of a live audience, three different authors, whom I would choose, would read aloud essays they'd written about the moral dilemma of eating meat. After the readings, one of Lisa's clients—a chef who knew how to butcher—would break down a side of beef or pig in front of the audience, and the farmer who raised the animal would share his or her farming practices. For the finale, we'd offer a parallel tasting of, say, grass-fed and grain-fed beef, or pastured and barn-raised pork. We'd pour wine, too, which, we hoped, would ensure heated discussion.

If you live in a city like Portland or Berkeley or New York, and you count yourself even remotely part of the food scene in that city, this sort of event is, today, probably not that novel. (Of course, anyone in a rural community who's reading this may be rolling their eyes.) But in 2009, Livestock was an anomaly in Portland. Tickets sold out within twenty-four hours.

"Livestock," we wrote in our press release, "is an urban conversation designed to explore the literary and literal aspects of killing dinner." John Berger would have loved it.

FOR THE FIRST NIGHT of the series, in mid-November, just a week or so after I returned from Hawaii, I decided to wear tight gray jeans and Sigerson Morrison heels from my days in New York. This was a conscious decision. I was going to be co-host of the event, and I wanted to strategically convey a contradictory impression.

Since I returned from France, I'd become increasingly aware of the way in which people spoke of me in relation to butchery, as if I were a monkey on display in a cage performing tricks no human had ever seen a monkey do: "This is my friend Camas. She just went to France to become a butcher. Can you believe it?" Or "You are one sexy butcher," even though I wasn't even really a butcher. "I'll make sure to never get into bed with you and a sharp knife," said a man I had only just met at a party.

I was pretty sure no man who'd gone to France to study butchery would be talked to this way. Underneath such comments I sensed a deep-set doubt that a woman who looked like me—whatever that meant—could ever be a butcher. Of course, I wasn't a butcher, not yet. But I had ambitions to be one, and I felt that, within all the joking about sexy butchers and sharp knives in bed, my very ambition was in question.

I figured I'd use this to my advantage. Those tight jeans and black heels would ensure that everyone made a certain judgment about me when they saw me. I wasn't sure what the judgment would be, but I knew it would not be *She looks like a butcher,* or even *She looks like someone who knows a lot about killing animals for food.* Then, when I spoke, I would obliterate their assumptions. I have always been a confident public speaker. I didn't know everything there was to know about meat, by a long shot, but I was pretty sure I already knew a lot more than most people in the audience, and I knew how to tell a good story. No one would see me wield a knife that evening—thank goodness—but they'd hear me talk about what it had been like for me to do so in France. I'd tell the story of my time in Gascony, and they'd think: *She* went and did *that?* They'd hear me unveil this idea for the Portland Meat Collective, and then they would ask me how to sign up.

Perhaps most important, I would put a face (and a body?) on butchery that was new. A so-called feminine face—attached to a quick-thinking, articulate brain. And this, I hoped, would get people to shift their perspective so that they'd become curious about something that, until then, had been the territory of intimidating, burly dudes in chain mail and XXL white butcher's coats, with blood and tattoos on their forearms, who didn't much like to speak to outsiders and held the nature of their work close to their chests.

Jo showed up the first night. I hadn't called her yet. When she walked into the room, her presence immediately unnerved me and so I turned away before our eyes met. She sat just to the right of where I was to stand the entire night while I emceed the event.

We'd crammed an audience of fifty into a kitchen at a local culinary school. No one seemed to mind the tight quarters. I welcomed everyone to the event and then announced our first writer, who read an essay about the pleasure she felt eating meat, even though she felt guilty eating it.

"I am a lapsed vegetarian, the way my father is a lapsed Catholic. But while my father lost his faith in rosaries and Hail Marys, I've simply been ignoring my belief in the essential rightness of vegetarianism. My dad will never stand in a Communion line again, but one of these days I'm going to return to a plant-based diet. I've just been putting it off."

The next writer read a story about going deer hunting with her father after he'd been diagnosed with a particularly virulent form of brain cancer.

"That day in the woods, after the deer was field-dressed, my dad turned to me and he said, 'Well, that was maybe my last deer.' I bent a little farther over the knives as I cleaned them, trying not to let him see how close I was to crying, and he picked up his camera and snapped my photo. But he didn't need to. There was a deer. I was with my father. It's a story. For as long as I can, I'll remember."

Another told the audience about his experience watching a butcher kill live eels in front of a massive Buddhist temple in Japan.

"Wriggling and wet, the eels received a spike through the eye, and were appended onto a bloody block of wood by the ocular point."

Then our butcher for the evening, Adam Sappington, set to work on a massive forequarter of beef. Adam was a longtime Portland chef who'd started butchering whole animals in restaurants nearly a decade before *The New York Times* dubbed butchers the new rock stars. He was a good ol' boy from Missouri who, with his wife, Jackie, had just opened his own restaurant, the Country Cat, which offered Southern dishes like fried chicken but also Pacific Northwest fare that James Beard would have approved of, like pan-seared salmon and fried oysters. Each week he ordered a whole lamb, an entire side of beef, a side or two of pork, and dozens of whole chickens and broke them down himself in his diminutive kitchen. All the animals came from local, small- to medium-size farms, and he'd somehow figured out how to use every part of each animal on his menu *and* make money. Few chefs were doing this in Portland at the time, and if they were, it was usually as a novelty, for an occasional special dinner or menu. Adam had figured out his own narrow path across an indeterminate expanse of known and unknown risks. He didn't own his own farm, but he'd inserted himself into America's food system and figured out his own defiant workaround.

I became completely disoriented watching him butcher that side of beef. His particular reading of the road map, his technique, bore little resemblance to what I'd seen the Chapolards do. Sure, it was beef, but weren't the parts of the animals mostly the same? I momentarily wondered if everything I'd learned in France had been totally wrong.

But then Adam told the audience that he'd taught himself how to do this, that he'd figured it out based on the cuts he needed to please the customers who paid his bills. And the parts his customers believed they didn't want to eat? He'd disguise them in the form of stocks and broths, savory pies, or "bacon bits" sprinkled over salads. Of course he cut the animal differently. Plus, it *was* beef. What did I know about beef? Nothing, really.

Afterward, we served each of our guests composed plates of slow-cooked grass-fed and grain-fed beef chuck and slices of lightly seasoned and roasted beef loin. The audience tasted each and told us what they thought.

"Too meaty," many people said of the grass-fed beef.

"Too fatty," others said of the grain-fed beef.

"I can't even tell the difference," one guy admitted.

The farmers who'd provided the meat entered into a friendly debate about grass-fed versus grain-fed beef. The grain-fed-beef farmer said she'd consider grass-fed if she had enough land to pasture her animals on and if consumers actually liked the taste. The grass-fed-beef farmer insisted it was the only way to raise beef, that it was better for the environment and the animals, not to mention better for our health, that educating consumers about the unique flavor was part of her job, that consumers would have to open their minds if they wanted our system of meat production to change.

I asked a lot of questions that made the farmers straighten up a little bit in their chairs. At what age do you slaughter your animals? Have you ever considered slaughtering them at an older age? Why not? What breed do you raise? Why? How much do your animals move around in a day? I couldn't tell whether my line of questioning made the farmers happy or nervous, but I was pretty sure they weren't used to being asked so many questions.

As people filtered out, Jo grabbed me by the elbow.

"Hey, lady. I been hoping you'd call me. I wanna buy you some whiskey and pick your brain and help you start this meat school."

I had her write down her number again, just in case I'd lost it, although I knew exactly where I'd tucked it into my wallet.

She looked me up and down before she left. "Nice shoes, by the way. Good choice."

After cleanup, I asked Adam if he'd teach my first Portland Meat Collective pig butchery class.

"Sounds good to me, doll. Come over to my restaurant and we can talk over a glass of whiskey and a plate of beef jerky."

How was it that these people kept landing right in front of me exactly when I needed them? Two Southern butchers who called me "lady" and "doll," who wanted to help however they could and talk meat over whiskey and beef jerky, who, just like Kate, didn't even flinch when I told them my plans?

Tout seul, tu meurs, Dominique had said.

My idea began to feel a lot less improbable.

TWENTY-EIGHT

In the middle of the Livestock series, Andrew called and asked me to dinner. He took me to an overpriced surf-and-turf restaurant on a hill overlooking the city. I was grateful for the meal, of course, but when he asked me what I thought of the food, I couldn't help but point out that the lobster had freezer burn and the steak probably came from a feedlot in Iowa. He laughed, not in any way offended by my directness.

"Why don't you make all the food decisions from now on," he said.

"Great," I said. "Have you ever tasted raw sea urchin before?" He hadn't, but he was game to try.

We were, in so many ways, nothing alike. Andrew had grown up in the flat expanse of Illinois cornfields, amid waving grain, Chicago sky-scrapers, Republicans, football, fraternities. I was named after a wild-flower and a ditch and dropped acid for the first time at the age of fifteen, the same age I decided tempeh and tofu were more righteous than my dad's venison and trout. By the age of sixteen I'd formed a feminist club at school and volunteered on the weekends to unionize migrant farmworkers in Woodburn, Oregon. I came from a family of liberals, maybe even radical liberals, at least in the eyes of the people Andrew grew up with—but with a certain rural flair, a rural flair that Andrew actually did share with me, having grown up on a

Christmas-tree farm with horses. Both of us had a little country in us, but my kind of country now included the word *charcuterie*, which Andrew wasn't yet sure how to pronounce. Our backgrounds were so unknowable to each other that we began with a remarkable lack of assumptions about each other.

A FEW WEEKS into my new romance with Andrew, I left a voice mail for Jo. "Hi, Jo. We met at your butcher shop? I'm the one starting the meat school? It'd be great to meet up and chat. Seems like we have mutual interests."

"Mutual interests" was an understatement. What I really wanted to say was: *I feel so lucky to have found you. It feels lonely to be me right now, a woman out here in meat land, asking questions I'm not supposed to ask.*

We met up at a new Spanish restaurant in town and shared a plate of foie-gras-stuffed dates and a burger, not to mention several tiny glasses of whiskey and beer—tiny glasses that we emptied quickly and that were refilled even quicker.

I told her more about the Portland Meat Collective and she immediately offered to help. I confessed that, even with her help, I wasn't sure I could pull it off. Things seemed to fall apart in my life, I told her, and I had little money to my name. I told her about Tom and Will, about losing my job, and she responded by telling me that she'd recently lost her high-paying corporate job before deciding to work her way into the world of butchery. She asked me about France, and I told her how it had changed everything for me.

"That doesn't sound like falling apart," she said.

"I just want to keep learning," I said. "And there's nowhere to learn."

"Me, too," Jo said.

"But you can learn at the butcher shop," I said.

"Not really. I mean, Matt and Stu teach me things every once in a while, but they're burned out. They've worked there for over ten years. Besides, we don't even do that much whole-animal butchery. I mostly just make a lot of sausage. Let's make a place where we can learn," she said.

It might have been all the whiskey, but talking to Jo over those little glasses of liquor, I began to wonder whether the universe was more magical and all knowing than I had ever believed, having sent this remarkable woman to me, a woman who, save for her Southern accent and my hunch that she slept exclusively with women, appeared to have been living a near-mirror image of my life as of late. Jo offered to pay our entire bill, and then we made our way upstairs to a tiny bar and ordered more whiskeys.

Five whiskeys in, we were making out at the bar like teenagers. What in the hell was I doing? Was I even attracted to Jo? Was this even me? And what about Andrew? I liked Andrew. A lot. I liked the way he came into my life sideways, from unfamiliar, unknowable places. I liked the mystery in that. Jo, on the other hand, felt so familiar in the most uncanny of ways. How could she possibly exist? The writer in me felt suspicious. Seriously? I'm the only woman I know who's left her career to learn butchery, and Jo goes and does the same thing and then we meet over a stack of rib eyes? Was I about to have an affair with a lady butcher? *If I ever write this story, no one will ever believe me.* That's what I thought while making out at the bar with Jo.

"I'm seeing someone," I told her. But we headed to her house anyway.

I left early the next morning, feeling guilty and hungover, unsure of who I'd woken up as, or who I'd woken up with, but I didn't have long to process any of it. A few days later, Jo showed up at my place with butcher paper and knives. I stood at the door, and didn't invite her in.

"You gonna let me in?" she said. "Let's start this thing. I brought paper and pencils so we can sketch out a business plan. You ever written one before? I'm pretty good at it." She pushed past me into my tiny apartment and flopped onto the couch. I stayed by the door, as if by sticking close to that particular threshold I might, at any second, be able to pass through it and go back in time.

"About the other night," I said. "I don't think we can do it again." I couldn't say exactly why. There was Andrew, of course, but we'd made no specific commitment to each other yet. Maybe it was the fact that I'd failed miserably at dating women in college. Or maybe I felt we were meant to achieve more important things together. Maybe I was just scared. Maybe I knew I couldn't ever be for her what she needed me to be. We were so very much alike, and yet our meeting seemed so improbable—too good to be true, really. Better to cut this off now than to risk more heartache.

"If you just want to be friends, you can tell me. It'll be fine," she said.

"It's probably best," I said.

BUT THAT DIDN'T WORK. These were two incomparable loves. I loved a woman who could nearly read my mind. I loved a man whose mind I could not immediately know. For a while I convinced myself it would be perfectly acceptable to cultivate both at the same time.

This was something other people—worse people than me—did. Driving across town late at night, stealing from one person's bed to the other. This was the territory of liars and cheats, the subject of racy novels and lurid soap operas. This wasn't my wheelhouse. It never had been.

Jill called it "unleashing my awesome power."

"Just embrace it," she said. "It's complicated, but so what? What in life isn't?"

I'd returned from France ready to trade in my life of *buts* for *ands*. I hadn't planned on applying that to my love life as well, and yet I wanted to live in a world where I was allowed to love both Andrew and Jo in whatever way I wanted to, without consequences.

This is the most humane way to kill a pig, Marc Chapolard had told me. *You commit to the act. You do it fast. You render her senseless to pain. And then you bleed her. Any other way is just cruel*, he'd said. *Don't draw it out.*

There's another side to that equation, one the Chapolards never felt the need to tell me, perhaps because it seemed so very obvious to them. Even when you commit to the act, and the resulting sacrifice, even when you do it fast and do it well, you yourself do not become exempt from feelings of discomfort and pain. There's still sacrifice and sadness and discomfort to contend with.

I had become so good at pretending I was exempt from such complexities. And so, to stave off the inevitable pain, I did my best to close the door on Jo in my mind when I was with Andrew, and to close the door on Andrew in my mind when I was with Jo.

I have never been fully able to reconcile all that was gained and all that was lost when I finally made my choice. The right thing to do never quite felt like the right thing to do.

I'm talking about love and heartbreak and slaughter and dinner all in the same breath. I don't think I am supposed to, but I'll keep doing it anyway.

TWENTY-NINE

Two months or so into my messy entanglements with Andrew and Jo, while Jo and I also worked together to make the Portland Meat Collective a reality, an article came out in the state newspaper detailing my trajectory from magazine editor to aspiring "ethical butcher," as they called me. In the article, I announced that our first classes would take place in February, even though I still had so much to figure out. The day the article came out, several hundred people signed up for the Portland Meat Collective mailing list. I was in business, sort of, except I didn't actually have the money I needed to start the thing. So, I conceived a risky plan to require all students to prepay for each of my classes, then I could buy the tools and supplies I needed to pull it all off. Since the article made me sound more official than I actually was, I hoped that this rather rickety business plan, along with the kindness and generosity of people like Jo and Adam, would work. The article also transformed me, overnight, into a sought-after public speaker on the matter of the ethics of eating meat, a topic I was only just beginning to articulate for myself.

A few days after the article appeared, a professor at Portland State University called to invite me to take part in a roundtable talk. Jonathan Safran Foer had just come out with his book *Eating Animals*, and the talk would have the same name. The professor had attended the Livestock event, and he thought it all brought up a lot of sticky issues worth delving deeper into. For the talk, I would be charged with addressing "ethical meat eating," although this was not a term I had yet used to describe my own still-forming philosophy. A philosophy professor would give a ten-minute history lesson on the trajectory of moral and ethical thought about eating animals, from Pythagoras to Peter Singer. And a law professor would discuss the legal rights of animals. After we each spoke, we'd field audience questions.

Our moderator posed the challenge of the evening's theme to us this way: "If we all agree that we must eat, and eat well, then the really interesting questions begin when we ask: How do we eat well? And within this apparently simple, but actually massive, question, the issue I hope we can flesh out is: How can we eat animals well, if at all?"

I liked the question precisely because of its complexity, because of the way it inherently challenged the dogmatic thinking that so often accompanies discussions about eating meat. It was a question I wanted to answer, though I wasn't sure whether I was up for the task or not.

THE NIGHT OF THE TALK, my heart beat with a limp. My hands shook when I picked up a glass of water. I hadn't been able to decide what to wear, but since my armpits refused to quit pouring sweat, I finally settled on black jeans and the black turtleneck I'd so often worn in the Chapolards' cutting room. I'd just recovered from a severe case of pneumonia and possessed the darkest of circles under my eyes and the palest of skin.

"You look a little bit like death," Andrew joked. "It's very appropriate to the subject matter."

"I'm so nervous," I admitted to him before leaving his house to drive to the event early, alone. "I'm never nervous about these things."

"You'll kill it. You always do," he said.

It didn't help that five minutes before the talk began, Andrew walked in, followed by Tom and his new girlfriend.

From the front of the room I watched Tom and Andrew—who knew each other through our mutual friends Hava and Scott—shake hands and exchange a few words, their gestures appearing to me overly formal and forced. As they headed toward their seats, I watched Tom's new girlfriend crawl over a couple of people she appeared to know. She wore her long hair in two braids down to her butt. The thick green-and-black wool Pendleton shirt she was wearing expressed an outdoor West Coast ruggedness, the sort of look I'd cultivated back when Tom and I met in Eugene. It had been almost two years since he and I had ended things. While we'd mostly managed to remain friends, the pain was still raw so seeing Tom with this younger, more outdoorsy version of what he had probably always wanted me to be turned my skin hot and itchy with unwelcome jealousy.

Jo snuck in at the last minute, wearing a fedora low over her eyes. She stood in the back, behind a tall brick wall of a man, so that I could barely see her. I wondered if she didn't want me to know she was there. I wasn't sure I wanted to know. Tom and Pendleton Girl and Andrew were enough to contend with.

I was about to stand in front of a roomful of people to talk about the importance of transparency. I was about to try to encapsulate, in ten minutes, the nuanced, controversial nature of multiple competing truths: death, life, dinner. Meanwhile, my own personal competing truths were floating out there in the audience, their eyes on me. What knowledge did I possibly have to offer any of these people?

Pythagoras was a vegetarian, the philosophy professor began, and Aristotle thought hard about the moral status of animals, too, she said. She conjured Aquinas, Descartes, Voltaire, Tolstoy, Bentham, John Stuart Mill. She talked about the "negative rights" of all animals—that animals, like humans, had the right to bodily integrity, a notion I agreed with, although I was perfectly aware that many people believed that killing an animal for food, no matter how humanely, violated an animal's right to bodily integrity. She moved on to Peter Singer, the godfather of the modern animal rights movement, telling us that Singer believed we should seek to minimize suffering and maximize well-being and happiness. I also agreed with Singer. Then she turned to Kant, who generally believed it was acceptable to raise animals for meat, but unacceptable to be cruel toward animals. Certain Kantians, she suggested, believe we have a direct duty to humanity and an indirect duty to animals to not treat animals badly, because when we treat animals badly, we are, in turn, adversely affecting ourselves. I got a little lost in her particular rendering of this argument, but felt I could agree with its basic tenets, too. The philosophy professor didn't say whether she ate meat or not, but I sensed she was very much against it.

Next, the animal-law professor introduced herself, acknowledging that she was a twenty-year vegan. "Philosophers might believe that animals should have rights, but when it comes to legal rights in this country, animals don't have any," she told us over the top of her round Gandhi glasses. "I come to this issue from a social justice point of view," she revealed.

"Here's how my lawyer logic works," she said. "How many people agree with the following statements . . .

"We are members of the animal kingdom. Raise your hand if you agree." All hands raised.

"We have a right to survive." All hands raised.

"Eating is part of that right." All hands raised.

"We shouldn't cause unnecessary suffering." All hands raised.

"Animals can suffer." All hands raised.

"How many of you believe that animals suffer in the process of be-ing raised and killed for food?" Most hands raised. I agreed that most animals did, but I didn't agree that all animals had to.

"How many of you believe that we don't need to eat animals in order to live healthful lives?" Hands all the way up. Hands all the way down. Hands wavering. One guy groaned and then got up and left. I wasn't sure yet what my own answer to this part of the equation was, but I had begun to look into what happens to the body and the brain when animal fats and protein are cut out completely, and I had encountered an equal number of convenient nutritional narratives supporting the total aboli-tion of meat from our diet as I did narratives that supported some meat, in moderation, as beneficial. No scientist or nutritionist I could find could prove that the way Americans currently eat meat was essential to our well-being. At any rate, this didn't matter, because there was clearly no room for nuance in this lawyer's argument.

"My conclusion, then," she said, "is that killing animals for food that we don't require for survival is unnecessary suffering and is there-fore without moral or logical authority." Logical indeed. Her equation seemed so very simple. I wasn't convinced.

When my turn came to speak, I wanted to know whom I was speak-ing to.

"How many of you are vegetarian or vegan?" Three-quarters of the room raised their hands.

My talk, punctuated by a good many *umms*, went along different lines than the first two speakers. I wasn't going to talk about whether or not we should eat animals, I told the audience. I was more curious whether, for those who choose to eat animals, a middle ground exists between choosing not to eat them at all and choosing to eat animals at the cost of their suffering. What were the options for folks who felt they wanted or needed to eat some meat in order to live a "healthful life" but raised their hands for the rest of the lawyer's questions? What were the options for people who understood that, more often than not, raising animals for food did cause suffering but did not always have to? What if you could raise an animal for food and kill it without any excess pain or suffering outside of the normal pain and suffering that all animals, and humans, can expect to encounter as sentient beings living in a volatile and unpredictable world? The only sticky question we'd be left with then would be whether killing an animal—absent any pain—was moral or not, and that seemed to me a largely personal choice that necessarily had to be informed by your relationship to the *ands* and *buts* of the world.

The audience blinked back at me. I coughed. The *umms* gave me time to breathe and think between ideas. The *umms* probably also made me a shifty suspect.

And then our even-tempered, fair moderator let the audience out of its cage. He could not contain them.

"Please raise your hand if you have a question for any of our speakers."

The people in the audience who raised their hands—and there were so many—mostly delivered speeches of their own. Indeed, some seemed to have practiced in front of the mirror beforehand. What questions they did ask were mostly aimed at me.

"Do you consider yourself a feminist?"

"Yes," I answered.

"Then how can you eat meat?" The audience member proceeded

to pull out a copy of Carol Adams's *The Sexual Politics of Meat* and wave it at me.

"I've read the book," I told her. I'd actually read it three times in college and still had my dog-eared, highlighted copy at home. Nevertheless, she began reading from the book.

"An integral part of autonomous female identity may be vegetarianism; it is a rebellion against dominant culture whether or not it is stated as a rebellion against male structures. It resists the structure of the absent referent, which renders both women and animals as objects."

"Ma'am," our moderator said politely, "we'd like for this to be a time for questions. Moving on. You, sir . . ."

A gray-haired man in tie-dye stood up and looked at me. "Do you consider yourself a Nazi?" He pointed his finger at me when he said it.

"No," I said.

"Well, since you believe in imprisoning animals, you're a Nazi to me."

"And a slave owner!" someone in the audience added.

"As a lesbian and a vegan, your speech actions violate me," one woman said, her eyes locked on mine. I'd watched her during my talk as she shook her head often, pursed her lips, and struggled to stay seated.

I glanced over to where Jo had been standing, but she'd already left.

The very last person in the audience to voice his opinion told us he'd grown up in Turkey. He grew up eating meat, he explained, although not very often, because it was a special privilege and hard to come by. But he reminded us that, for poorer nations like his, eating meat was and still is a matter of survival.

"This is a very American discussion," he said. "For people elsewhere, we don't have the luxury of such debates."

"And on that note," our moderator said, sounding a little exasperated, "thank you all for coming. We have animal crackers for you to enjoy before you leave."

I SAID HELLO TO TOM, shook his new girlfriend's hand. She told me she didn't eat meat but she had liked how I framed the debate. She also asked me if I ever taught rabbit classes and said that if I did, she'd take all the pelts—for an art project, she said.

The woman with the *Sexual Politics of Meat* book pushed past me to the door in a huff.

An anthropology professor from Lewis & Clark College asked me if I would speak at one of her classes.

A farmer asked me if I'd buy her pigs.

One woman handed me a PETA flyer that said MEAT IS MURDER.

Andrew and I drove in separate cars back to his house.

On the way, Jo texted me.

"You killed it!" she said.

I wasn't so sure I had killed it. I hadn't been prepared for all the pushback.

AT ANDREW'S HOUSE, I crawled into bed with him, nestled my head into his chest, and shut my eyes.

"I have to tell you something," I said.

Seeing Andrew and Jo in the same room had forced both doors in my brain to open to each other. I had known, for some time, that it would be impossible to keep the two separated both in the real world and in my head, and that once any semblance of a collision occurred,

I would be forced to contend with what felt like an inevitable decision. Jo already knew about Andrew. It was time for Andrew to find out about Jo.

Andrew got out of bed, pulled on a shirt and pants, and began pacing back and forth. "We never made any commitment to each other," he said, grabbing a pack of cigarettes and a lighter and walking out of the room. "You have the right to see someone else, but still," he said, shutting the door loudly behind him.

But. Still.

Our chances for survival were so very slim, really, but somehow chance won out. That would be the easy, romantic story to tell, anyway. The truth is, those first months together with Andrew were wild and rough and untethered and wholly uncertain, and I was almost solely at fault. I fell in love in two very different ways with two very different people at once. For such a long time, I was unwilling to lose one to keep the other. The real, practical world needed me to make a choice. I wanted to believe I didn't have to. *That* was where the fault lay.

When he came back inside, I told him I'd call it off with Jo. And the next day I did, again, sort of, except not really, because neither of us knew how to live within the absence. And so, over the next several months, Jo and I chose to separate ourselves from each other in the slowest way possible, constantly second-guessing whether it was the right decision, struggling to redefine who we would be toward each other, searching, always, for the right vocabulary to capture the loss we both felt. I could not explain this to Andrew, so I didn't, not for years, not even after Jo and I almost completely stopped speaking to each other. When I finally did, I still couldn't find the right words.

THIRTY

One of the reasons I drew things out with Jo for so long was that I couldn't have started the Portland Meat Collective without her. Even in the middle of our knotted affair, Jo relentlessly pushed me forward. She introduced me to farmers, took me to slaughterhouses, brought me knife shopping. Our shared enthusiasm was sincere and contagious. We felt we could do great things together. We knew all the right people. We had exactly the right combination of skills. Together we were charming and funny and singular in our way, and people were drawn to us—these two women who didn't belong, who'd chosen to do something utterly unexpected. And then we were ready for our first class.

For the class, at the recommendation of Adam and Jo, I'd ordered two pigs from Mile End Farm, which sold whole and half pigs and smaller cuts like pork shoulder and belly to several restaurants in town, including Adam's, as well as to meat counters like Jo's. The farmer, Tricia, was older and had a dry, ashy voice. She held back her frizzy hair, dyed a burnt golden red, with a big 1980s-style butterfly clip. And she carried around a notebook with pictures of all of her pigs, whom she referred to by name. She told me that in winter she let some of them sleep indoors with her. I wasn't sure whether to believe her or not.

"I love my piggies," she'd coo, hugging her notebook to her chest.

She didn't pasture her pigs, she told me, but they lived in an open-air barn that sounded kind of like the Chapolards', with plenty of room to move around.

I hadn't yet found a farmer who pastured *and* raised enough to be able to sell to me at the last minute (and my first pig order was very last-minute), so she would have to do, even if I hadn't yet been able to visit her farm myself. Plus, it hadn't occurred to me not to take the farmer at her word.

"No problem. I can get them for you on Friday," she said to me the Monday before. This probably should have aroused suspicion, but I did not yet know which red flags to look for.

Adam asked the slaughterhouse to keep the two carcasses I had ordered whole—"luau style"—which is how he'd taught himself to butcher animals. This request tacked on an extra hundred dollars per animal for some reason, but the animals themselves were remarkably inexpensive, at $1.67 per pound, and the pigs were small, with hanging weights (meaning their weight after they are slaughtered, bled, and eviscerated) of about 150 pounds, less than half the size of the Chapolards' pigs. At the time, this very low price did not seem odd, either, not yet.

The class took place on a Sunday at Zenger Farm, an urban farm and educational nonprofit at the very eastern edges of Portland's city boundaries that held summer farming camps for kids, hosted farm-to-table dinners, and offered community-supported shares in all of its produce. In the farmhouse kitchen, there were rickety folding tables for us to work on and a counter with a stove in the middle of it that we covered up with Adam's oversize cutting board. A regular household fridge had already been filled with the employees' lunch leftovers, so we would have to use the big coolers that Jo brought to keep all the meat cold.

Eight students had signed up. They varied in age and were about half men and half women. While Jo and Adam greeted everyone in their warm, Southern manner, I busied myself with last-minute details: slicing charcuterie for the end of the evening, counting knives, sanitizing cutting boards.

The three of us introduced ourselves to the students, then we went around the room to find out why they'd each signed up, something we have done at every class since. There was a bike messenger, an IT guy, a lawyer, a product designer, a hunter, a college student, an aspiring farmer or two. Some wanted to eat only meat that came from animals they'd raised themselves. A few just wanted to know how to buy better meat from local sources. A few of them wanted to occasionally start buying whole animals from local farmers and butcher them on their kitchen counters once or twice a year. Everyone wanted to know why bacon tasted so good.

Then we got to work. Again, I felt completely lost watching Adam. What was he doing with that cleaver and that mallet? Why did he cut the ribs and belly like that?

"I cut these pork chops this way," Adam explained, "because I have a whole-hog plate on my menu and one of the items on the plate is what I call a Flintstone chop, with the entire rib bone attached to the loin chop." It sounded a little excessive, and very American. It was also the kind of pork chop I might actually like—all that dark, fatty, flavorful rib meat probably made up for the lean, mild-tasting loin.

He showed us how he removes the pelvic bone, or aitchbone, from the back leg, followed by the shank and trotter, then walked us through his process for making a true American country ham. First, he buries the whole ham in salt, sugar, molasses, juniper, black pepper, and sage—adding a generous dash or two of bourbon, of course—and allows it to sit in that cure for two months under refrigeration. When he feels it's ready, he soaks the leg overnight in water and the next day he hot-smokes it over hickory for twelve hours.

"But for this class, we're going to break down the hams into separate muscles," I said. I shared with the students how the Chapolards made their tiny salted, smoked *jambons* and suggested that this might be a less risky way to start learning about the principles of salting and drying meat.

"Each of these muscles has a different name depending on who you're talking to," Jo said. "At my shop, we call this gnarly muscle by the kneecap the baseball roast, but I've heard other butchers call it the knuckle."

"I just call it ham," Adam said.

The students nodded and pointed and asked questions, wholly absorbed. They wondered aloud if they could get these lesser-known cuts in butcher shops, or from farmers at the farmers' markets.

"If you start asking for them, I bet you can," I said.

Once Adam finished his demonstration, we had the students put all the pieces back together again. They mistook the shoulder for the ham, just as I had in France, but mostly they figured it out.

We brought everyone over to one of the low, rickety tables where we'd set the other pig.

"Go slow, ask a lot of questions, try not to cut yourself," I said. "But if you do, we have Band-Aids."

The students each picked up a knife and stared at the pig.

"Not all at once, folks. Take turns," Adam said. "One person take the trotter off. Then the next person removes the hock, and the next person cuts the shoulder primal off of the belly and loin."

"Remember your butcher's grip," Jo said. "Your wrist won't hurt at the end of the day, and you'll be more precise."

Thus far, no one had made a psycho killer joke.

The students learned to guide the ends of their knives through fascia. They sawed through bone. Their cuts were jagged at times, just like my first cuts had been and probably still were. But the chops mostly looked like chops. The gangly roasts would glue themselves back together once they were put in the oven.

I told the students how different this animal looked from the pigs I'd worked on in France. The meat was more pale, a bit soft and watery, with very little fat. I didn't know enough to be able to tell if this was due to the handling of the pig before slaughter or whether it was the breed or the feed or the age of the pig, or all of the above, but I wanted them to know there was a difference.

"We're not typically raising pigs for charcuterie in America. We're raising them for lean-muscle cuts," I said.

After three hours or so, we'd broken down all the meat together. We poured them wine and passed out charcuterie and the students began to talk to one another, expressing concern over how most meat they ate was raised, wondering out loud how they might buy better meat and who the farmers were who raised that meat.

When the students left, each of them telling me they'd sign up for whatever class was next, we had about three hundred pounds of meat, fat, skin, and bone to wrap. We'd told the students that we'd split it all up as evenly as possible among eight people and have bags ready for them to pick up the next day.

Adam pulled a six-pack of beer out of his bag. "We'll need this."

We stayed for another two hours, wrapping each cut of meat in butcher paper, weighing every package, labeling it with a Sharpie, and stamping it with a NOT FOR SALE stamp, which the USDA required. Given the nontraditional nature of our students having cut it themselves, the USDA didn't want this meat to enter into the commercial food system. It was now deemed unsafe, except for the people going home with it.

"What should I call these, Adam? Loin chops? Rib chops?"

"Flintstones!"

Between setup, cleanup, and the class itself, we'd worked about ten hours, and Jo and I still had to transport all the meat to Pastaworks, where we would split it all up into paper bags for each student and then store it in their walk-in until the students came to pick up their meat.

The learning curve was steep. I had no examples to work from. It would take us many more classes before we decided to use three sides of pig instead of two whole pigs—there would be less meat to contend with that way, and we wouldn't have to spend as much time splitting them into sides ourselves. After a few years, I'd price the classes such that I could pay assistants to help wrap all the meat during class and could afford to rent a more appropriate kitchen space with a walk-in so that we could easily keep the meat cold.

"Well, gals. That's it for me," Adam said. We helped him load up the back of his truck with his cutting board and tools.

"Let's do it again," I said.

"I'll be there, sugar," Adam said, blowing both of us a kiss.

THE SUN HAD SET by the time Jo and I were ready to load our two big coolers of wrapped meat into the large gray boat of a Suburban that Jo had borrowed from her friends.

"Honey, that was amazing!"

"Really? Was it? It was so much work. And what was up with those pigs?" I asked Jo. "They were so small and pale."

"That's what they always look like."

"We should go check out her farm," I said.

"Yeah. Maybe you are right. I'll call them and make it happen."

"Let's load these coolers."

We positioned ourselves at either end of one of the coolers and each gripped a handle.

"Ready?" Jo said. "One, two, three—go!"

We could barely lift the cooler.

"Let's try again. One, two, three—lift!"

We picked it up about an inch off the ground and shuffled it about two feet before setting it down again. I started laughing. We hadn't really thought through how much meat we were putting into one cooler.

"Okay. We can do this," Jo said, always the enthusiastic cheerleader. "Go!" We lifted and shuffled a few more feet, set it down, and laughed some more, and then did this about ten more times, until we got both coolers into the Suburban.

We parked in the alley behind Pastaworks, and again we pushed and pulled and lifted those coolers, two girls lugging three hundred pounds of hams and shoulder roasts and bellies and stew meat through mud puddles in a dark alleyway. We were out of breath, sweating, laughing, doubling over, tears streaming down our faces.

"This is some serious meat-mafia shit right here," Jo said. "If my mama could see me now."

"I don't think my mama would want to see me now," I said, still laughing.

"I can't think of anyone I'd rather be doing this with, though," she said, suddenly serious, wiping the tears from her eyes.

"Me neither."

I looked at Jo and wondered how I could possibly have made enough right turns in my life to find her there in front of me, lugging this cooler heavy with meat, heavy with our singular and entirely unlikely shared interest. I could not imagine pulling off any of this without her. We had, a month or so before, once again, decided to try to keep things platonic between us, but I wanted to press her against the wall right then and there and kiss her. "Let's get this done," I said.

We dragged the coolers through the back door to the kitchen. We split up every package of pork chop, belly, stew meat, and ham among eight bags. All the bones. Rolls of skin. Cubes of fat. I weighed and re-weighed the bags.

"Honey, you're crazy," Jo said. "No one is going to care if they didn't get exactly the same as everyone else. They will never know."

"I care."

"That's why I love ya, honey," she said. "But we've been working for twelve hours, so at some point we're going to have to get these bags in the walk-in and eat some dinner."

I looked down at the floor. Andrew was expecting me that night. He wanted to celebrate the success of my first class.

"I can't. I want to. But I can't."

"Not even for our first class? You can't even get a drink with me to celebrate? Jesus Christ, Camas."

"I already told Andrew I'd meet up with him."

"He's not the one dragging whole pigs around in alleyways with you," she said, trying to look angry instead of hurt. "Why does he get everything and I get nothing?"

"That's not how I mean it," I said. "It's not the same relationship. I can't give you both the same thing."

"I'll see you later." She walked out of the store and to her car, shaking her head.

THIRTY-ONE

Shortly after pulling off the first Portland Meat Collective class, Jo landed me a job with Pastaworks. Given our mounting tensions, it seemed like a terrible idea to work together, but I needed the money, and besides, it was the kind of job I had secretly longed for during my days as a restaurant reviewer.

Right before things went sideways at the city magazine, Pastaworks had opened up a tiny restaurant, Evoe, in a small space adjoining the store. Kevin Gibson, one of my favorite chefs in town at the time, billed Evoe (the name was from the opening invocation to Bacchus in Virgil's *Aeneid*) as an "enlightened European snack bar." Behind the long, tall wooden table where Kevin did much of his cooking and prep work, he had two burners, a griddle, an oven, and a sink to work with. On the wall behind Kevin, a vast chalkboard listed that week's plates, sandwiches, snacks, and soups, with wines, cider, and beer to accompany each. No more than eight customers at a time could sit across from Kevin at the table as he worked, with nothing separating him from his customers but jars of house-made pickles, bowls of Meyer lemons, baby artichokes, and fennel, a crock or two filled with wooden spoons, a stack of white plates. If there weren't seats at the table, customers had about twelve other seats in the whole place to choose from.

Evoe offered everything I would ever want to eat at just about any time of day. The menu reminded me so much of the food we ate at Kate's. Stinky cheese and charcuterie. Silky scrambled eggs and chanterelles on toast. Hand-cut slices of *jamón serrano*. Bowls of olives. Spiced almonds. Prosciutto-and-butter sandwiches. A salad of shaved raw delicata squash, toasted pumpkinseeds, fresh sheep's-milk cheese, and mint. Lamb meatballs and harissa. House-made merguez sweetened with paprika and cinnamon. Crispy duck breast topped with honeyed persimmons and peppery greens. The prices were reasonable, the ingredients simple, the flavors perfect in their purity and complex in their combinations.

I'd written a review of Evoe in the last weeks before losing my magazine job. I was required to visit the place only two or three times to review it, but I visited Evoe twice that. I just couldn't stay away. It was one of the few restaurants in town that did not trigger cynicism in me. The food felt honest and straightforward, the atmosphere humble, the room bright, light, and welcoming. I remember thinking how great it would be to work there.

That's the job Jo got me. Evoe needed someone two or three days a week to tend to customers and help prep and cook. It had been nearly two decades since I worked in a restaurant, and when I had, it was as a short-order cook on the line of a greasy hippie diner in Eugene and as a lowly sandwich maker at a surly East Coast deli. But Jo had introduced me to Peter, the owner of Pastaworks, to discuss using their space to occasionally store meat for our classes, so he knew who I was. At Jo's urging, I wrote and asked him to take a chance on me, playing up my food-writing background, my experience testing recipes. Kevin occasionally brought in whole animals, and he used meat from the butcher counter where Jo worked, so I also emphasized my recent studies in France and my continued studies through the Portland Meat Collective.

"Can you meet this Friday?" Peter wrote back.

I did an on-the-job tryout for one day, taking orders from customers, pouring wine, helping Kevin assemble sandwiches and peel potatoes for the next day's *tortilla española*. The job required me to do a little bit of everything—I washed dishes, scrambled eggs, took orders, filled people's water glasses, and boiled brine for pickles in between. I liked the idea of adding these tasks to the odd mix of work I'd already thrown together for myself: bookkeeping for my brother, writing when an interesting story assignment arose, organizing butchery classes. I stood. I sat. I cooked. I typed. I wrote. I did math. I dragged a knife along the curve of bone. I lugged three hundred pounds of meat down dark alleyways. I talked with strangers. I marketed. I spoke to the media. Among all of these tasks, I was utilizing every part of my brain and body.

"When are you going to get a real job?" my mom kept asking me.

"I like what I'm doing now," I told her. "I'm happy."

Of course, cobbling together so many different low-paying jobs didn't put me in the greatest financial position. Yet this stitching together of various jobs at their ragged seams felt like exactly what I needed to be doing.

KEVIN WAS A QUIET MAN, hard to read, prone to mood swings, but he was kind and patient with me. And because I was prone to mood swings myself, I had an easy time spotting his from a long way off. He knew I had a good knowledge base about food, but I still had to prove that my taste buds worked just as well behind the counter as they did as a diner at the table.

Kevin approached his ingredients thoughtfully, with an eye toward minimalism. "Rare, meditative precision" was the phrase I'd used in my review, but perhaps this had conjured the wrong image of him for my readers. His food was precise, but not in an overly fussy way. He wasn't meditative in the Zen-master sense of the word—in fact, I sensed he was a haunted man, and that the only way to keep from completely succumbing to the haunting was to cook himself as far away from whatever haunted him as possible. This, I often thought, was the best kind of cooking—cooking that saves your life.

"You did good," Kevin said at the end of my tryout as we counted our tips. Tips! I hadn't held that much cash in my hand for a long time. I felt rich.

"Not bad for a restaurant reviewer, eh? It feels funny to be on the other side now," I said. "I think I like it much better."

Writing about restaurants had required me to understand not only what an ingredient was, but what the chef had hoped to make it do. Now I not only had to understand what we wanted an ingredient to do, but I needed to help make it happen. I needed those scrambled eggs with chanterelles to be silky and soft, the chanterelles to be caramelized but still possess bite. It was one thing to say this about a dish, and another thing entirely to make it happen.

Kevin poured us two glasses of bright, effervescent Txakoli from Spain.

"So next time you come in, I'll teach you how to massage an octopus. Also, I have ducks, so let's make some of that duck prosciutto you told me about."

I'd landed my first restaurant job in fifteen years. I'd gone from managing editor of a magazine to water-pouring, dishwashing, scrambled-egg-making, marcona-almond-spicing, bus-tub-carting workhorse for one of Portland's most understated, talented chefs. I was even serving food to many of the chefs whose restaurants I'd once reviewed, along with other food critics in the city. I cared not one bit how it sounded or looked. I'd get a "real job" later. Or maybe never.

WORKING AT EVOE also meant I came into regular contact with many of Portland's farmers, who delivered carrots and squash and exotic greens—but also ducks, chickens, pork shoulder, and lamb—to our front door. And so Evoe became, for me, the perfect portal to the players of the Portland farm scene.

Some of the meat producers who came in were seasoned marketers, like John Neumeister of Cattail Creek, who had been selling lamb along the West Coast for decades. He had a loyal following and little trouble selling out of his product, but he'd worked hard for a long time to educate his customers, mostly restaurants and retail meat counters by this point. I began buying whole lambs from him for our lamb butchery classes, classes he'd often show up for, to introduce himself to our students and answer their questions. Occasionally I'd call him with a question like "So how is it that lambs can be grass-fed year-round if in the summer months the grasses dry up?" To which he'd answer, "The answer is long. Let's meet for coffee." Eight years later, I still buy lamb from him for our classes, and I am still learning from him.

Most of the meat farmers and ranchers, however, were newer to the game, eager to make an impression on Kevin and me. Many of these new farmers had spent most of their adult lives making good money in corporate jobs and then, quite recently, decided they were going to make a nostalgic go of the agricultural life, using their retirement money to buy land and raise grass-fed Angus cattle or pastured Berkshire pigs, free-range Red Ranger chickens or Muscovy ducks. So far as I could gather, they had figured out how to farm through trial and error and were convinced that they produced quality meat—in truth, the quality varied greatly—in a humane and sustainable manner, but few of them had counted on it being so hard to sell the meat they raised *and* cover their costs at the same time, let alone turn a profit. They'd conjured in their heads a fantasy customer base, one willing to pay higher prices for better meat, one that understood how much work and financial input went into their meat—a base like the one that bought our *saucisson* and *boudin noir* in France. But they'd quickly realized that even consumers who wanted to eat better meat seemed to believe that a pig was made up of only bacon and pork chops, that the breasts of truly free-range chickens would be as big as those of the factory-farmed variety, and that somehow, miraculously, higher quality would not translate to higher prices. Their customers were Americans who ate 265 pounds of meat a year and had grown used to paying very little money for that 265 pounds, thanks to the low and hidden costs of factory farming and the subsidization of grain.

Their customer base was also not, for the most part, composed of adventurous eaters. They were largely afraid of fat, flavor, and texture. They complained of the "gamy" flavors of grass-fed beef and scoffed at the "toughness" of pasture-raised chickens. And they favored cuts whose flavors and textures they deemed familiar—sausage, steak, burgers, chops, bacon, ham—leaving the farmers to contend with the fifth quarter, as Kate called it: the trotters and hocks, skin and liver. To recoup the cost of producing these parts they couldn't sell, the farmers had to increase their prices on the cuts that did sell, and the customers complained.

Their customers were also typically not creative cooks, or even adequate cooks. Often, the customers would cook stew meat from the shoulder, which required a longer cooking time with low heat, as if it were a loin chop, and then complain to the farmer that the meat had been tough. They had little understanding of how different muscles had to be treated differently in the kitchen.

Even more of a surprise to me was that the farmers themselves were often not creative cooks, either. They, too, admitted to not knowing what to do with a pig head or a trotter or a hock, so they were rather ineffective salespeople when it came to making every part of the animal sound appealing to customers.

Education seemed the only way to help these farmers, but the farmers told me they didn't have the time to do that sort of education themselves. They needed to sell their ham hocks today. If the Portland Meat Collective, however, could teach eaters to get excited about grass-fed beef fat and the flavor of liver, and chefs to get excited about making their own charcuterie and stock and sausage out of a whole animal, maybe someday we'd have that longed-for customer base. Maybe someday these farmers wouldn't be looking at me so desperately for an answer. It was, however, a long game.

And so, over the counter at Evoe, as I assembled fennel salads and toasted crusty bread, I watched the farmers hustling as best they could, and I wondered what they were going to do with all the pig heads they'd stored up in their freezers at the end of the month because no one wanted them. This, of course, was what inspired me to eventually begin holding pig-head butchery and charcuterie classes, but one class wasn't going to relieve the number of farmers I knew with freezers full of pig heads.

I thought about the Chapolards' vertically integrated business strategy, their cooperative ownership of equipment and slaughterhouse, the creative ways in which the family had figured out how to market and sell every part of their animals. I wondered what these farmers in Oregon, along with their customer base, might gain if I could bring Kate and the Chapolards to America to share their particular way of work and life. I imagined Dominique standing in front of a roomful of people in his beret, with his very French mustache, admonishing the audience that if you work alone, you die, and then telling them to stand up straight, breathe, and smile before demonstrating how to make a good head cheese.

I called Kate.

"Kate, it's Camas. How much money would it take to make a trip to Oregon with Dominique worth your time?"

THIRTY-TWO

As it happened, Kate was already planning a trip to Portland to attend a gathering of international culinary professionals in April 2010, and so we decided to pull together a workshop for the conference. We pictured it as a cross-cultural comparison of whole-animal utilization and butchery. Adam agreed to do a demo alongside Dominique, with Kate acting as Dominique's translator. To help us moderate, we also invited Michael Ruhlman, a prolific food writer who, in 2005, had teamed up with chef Brian Polcyn to write and publish the seminal meat-curing book *Charcuterie*, now a kitchen bible to many chefs and DIY charcuterie makers.

Kate had come to the States many times to teach, and Dominique had traveled to the States a few times before to visit his son, who for a short while lived and worked in Florida, but this would be Dominique's first time butchering in front of an audience of more than one or two people. Kate told me that his brothers had objected to the trip. It was a long time for him to be away from the cutting room and markets, and they didn't understand how Dominique teaching butchery to a bunch of Americans would ever benefit them. Dominique was able to convince them to let him do it only by agreeing to split any profits he made with them.

Dominique and Kate arrived a few days before the workshop, so I drove them around Portland to taste the charcuterie that a few restaurants and meat counters were producing at the time. At each stop, Dominique would pick up a piece of salami or ham or coppa, rub it with his fingers for a few seconds, and bring it to his nose to smell, before taking a bite. Then he'd chew for a while, look bewildered, and say, "I can't taste the meat."

"What does he mean?" I asked Kate. I'd taken them to the restaurants and meat counters I thought were making the best charcuterie in town.

"It's the nitrites," she said. "He says it's all he can taste." I felt a bit embarrassed that I hadn't noticed. But once Kate said this, I remembered the Chapolards' charcuterie, and how simple their list of ingredients was—just salt, pepper, fat, meat—yet how complex it was in flavor.

In America, most companies who commercially produce cured meats—be they of the fermented variety like salami or cured whole muscles like coppa or ham—include the preservatives sodium nitrite and/or sodium nitrate in their recipes. (These are sometimes referred to as curing salt, pink salt, or Prague powder.) These curing agents are typically commercially produced by adding small amounts of nitrite or nitrate to salt, although some producers use natural ingredients that are inherently rich in nitrates, such as celery powder, so that they can legally label their charcuterie "uncured." This is entirely misleading, though, since they are still curing their meat using nitrates, with the same intended outcomes. Regardless of whether natural or commercial forms are used, these preservatives have historically been used to inhibit the growth of the harmful bacteria that causes botulism—although the risk of botulism greatly decreased with the prevalence of refrigeration—but they also happen to extend shelf life, prevent rancidity, and keep the meat a rosy red or pink color, all important factors to be able to control in the modern commercial charcuterie world.

Given all of this, I asked Kate how the Chapolards got away with not using nitrites or nitrates in France.

She said it wasn't required of producers like the Chapolards and that it wasn't part of their culture—although, to be clear, there are plenty of charcuterie producers in France who choose to use these preservatives, and it's not technically required of producers in the States either. She also suggested that they were using the freshest meat possible. By the time they salted their *ventrèche* and *jambon*, only a few days (at most) had passed since slaughter, and they could vouch for every moment of meat handling and storage in between.

"The meat most chefs and butchers use in America is probably factory-farmed, and often slaughtered or processed several states over," Kate said. "Chefs probably don't know for sure how the meat was handled or under what environmental conditions they were transported or stored. If I were using that meat to make charcuterie I would use nitrite and nitrate, too."

I spoke with a meat scientist recently who has worked for some of the larger meat companies in America, and he disagreed with this, suggesting that a small butcher shop or a restaurant chef in America buying meat from a distributor would absolutely have that kind of assurance, he said, because everyone handling the meat along the way would be under strict regulations to handle it properly and keep it safe. I wanted to believe that everyone handling this kind of meat along the way would be sure to follow the rules, but given the amount of meat moving across our highways every day, and the number of people handling it, I wasn't sure I could muster total faith in that system of strict regulations.

Besides, Kate said, extended shelf life wasn't a desired outcome for the Chapolards. Their goal was to always sell out of everything they produced as soon as it was ready to be consumed, and they just about always met that goal. Kate went on, saying that when you have total control over the handling of the meat prior to curing, all that is needed to keep charcuterie safe to eat is the addition of the right amount of salt and the proper humidity and temperature for fermenting and drying conditions.

"Salt is a preservative, too," she said. "And it has been for thousands of years."

But without nitrites or nitrates, how had the Chapolards' *ventrèche* and *saucisson* still looked pink after weeks of drying and aging? That same meat scientist told me that it was entirely possible that the salt the Chapolards were using could be impure enough that it naturally contained its own nitrates, or that any water the meat came into contact with during the slaughter process could have contained nitrates as well. In addition, he said, the process of cold smoking introduces a lot of nitrogen dioxide, which can contribute to the pink color of the meat.

Dominique took another bite of the dry-cured coppa on the wooden board between us and said something to Kate in French.

"Even if this didn't have preservatives in it, he says he thinks the meat would have little flavor. He can tell the animals this meat came from were young," Kate said. "They hadn't developed much fat by the time they were slaughtered, and so the muscles were likely watery and underdeveloped."

"Does it taste bad?" I asked.

Dominique and Kate went back and forth in French for a while.

"Not bad, exactly," Kate said. "Just . . . American."

I saved our stop at Evoe for last. A month or so before, I'd finally coaxed more advanced instructions for Jehanne's foie-gras-stuffed duck prosciutto out of Kate, and Kevin and I had attempted to emulate it. The ducks we used hadn't been foie gras ducks, so they didn't possess the same amount of fat that Jehanne's birds did. These ducks had also been slaughtered at about half the age and weight of Jehanne's birds. But they'd come from an Oregon farmer who pastured his birds and slaughtered them himself, and we'd butchered and salted them the same week they were harvested. On the other hand, we'd had to order the foie gras from a local distributor who'd bought it from a producer in California. The foie gras had come to us vacuum-sealed, and we had no way of knowing how fresh it was. But when Kevin and I tasted our creation, we'd been mostly pleased with the results. Since we weren't planning on serving it to the public, we hadn't used any curing salts, just regular sea salt.

I sliced a few pieces for Dominique and Kate to try.

"Well?" I asked Kate.

"It tastes like it's supposed to on the surface. But you're missing all that nice fat that really drives the flavor of the recipe. Plus, how fresh was the liver? It doesn't taste as fresh as Jehanne's." Kate reminded me that Jehanne salted her foie gras and duck breasts the same day that she slaughtered her ducks.

"This is something we should talk about at our workshop," Kate said. "You can't just take the recipes from one culture and apply them to the ingredients of another culture. It's not a one-to-one ratio. There's an entire history and culture and method of raising animals that informed those recipes. The animals you raise don't necessarily make sense for our recipes. You guys are either going to have to raise animals differently, or you are going to have to come up with your own recipes."

It felt like a scolding, but it was an important one. It seemed to me that getting modern-day Portlanders, let alone Americans in general, to invent new recipes was going to be a lot harder than getting Americans to value (nay, crave) the recipes of another time and place, and then seek out and support a food system that would make those recipes taste good. If I could create a community of people—even a small one—who, say, wanted to make their own duck prosciutto and therefore began demanding fattier ducks, wouldn't farmers eventually change the way they raised their ducks? Wasn't that, in a way, its own network of narrow paths across the indeterminate environment of market forces in America?

IN LIGHT OF KATE'S SCOLDING, I wanted to bring her and Dominique to at least one farm that I felt *was* working toward the sort of animal production methods that Kate was talking about—even surpassing the Chapolards in some ways—methods that had the potential to serve the principles of those Gascon recipes well.

I'd chosen Square Peg Farm, about forty minutes away in Forest Grove, because I felt the owner, Chris Roehm, had a unique product—certified organic pork, fed locally sourced organic grains, raised on pasture. He'd forged loyal relationships with various restaurants, selling them whole and half pigs, or the equivalent in already butchered cuts, on a regular basis. And he also managed to sell a lot of cuts out of coolers at the farmers' market, convincing people to try parts of the animal they had never tried before, even providing them with cooking tips and recipes. In addition to raising and selling pork, he also grew organic produce, some of which he fed to the pigs.

As we snaked down the long gravel driveway to Chris's farm, we passed apple, cherry, and plum trees in bloom. Just as the Chapolards' property had looked like a true working farm, so did this one. There were plenty of outbuildings, sheds, and shacks. Wheelbarrows. Coiled hoses. I parked my car in front of a barn, and when we opened the car doors, we were greeted by the smell of wet Oregon soil, hay, and animals.

Chris, a ginger-bearded man with round, professorial glasses and a ropy build, who looked as though he spent more time growing food than eating it, emerged from one of the barns, smiling, although in that reserved manner characteristic of so many of the farmers I met.

"So this is it," he said after greeting us, opening his arms and turning in a circle to show us his farm's expanse. "We've got about forty acres. We rotate our vegetables on various fields over there," he said, pointing. "Over here we've got some fruit trees. We've got a few temporary hoop houses to grow greens in during the winter months. And right now, we've got our pigs in this barn." He pointed to a medium-size pole barn with a shed wall on one side and low temporary fencing making up the other "walls." We walked over and said hello to twenty or so adolescent pigs.

"We also rotate our pigs in the winter months into a few of those cold frames over there," Chris said, pointing to a trio of what looked like plastic-covered greenhouses standing out in one of the fields, with the shadows of pigs moving around within them.

"I thought you only pasture-raised your pigs," I said.

"We do, but it's really not possible this time of year."

Chris explained that for a few years he'd tried to truly pasture his pigs year-round, but in our wet and rainy Willamette Valley, he felt it just wasn't a viable option. His fields, he told us, were continual mud pits, where nothing could grow and erosion problems were common. Even with careful rotation, his farm wasn't quite big enough to sustain a fully pasture-raised, two-hundred-pigs-per-year operation *and* a vegetable farm. So he'd eventually settled on keeping them housed in the open-air barn and the temporary cold frames from October to April. During that time, Chris fed them organic grains he bought from another farmer, alfalfa hay, spent chestnuts from a gluten-free brewer in Portland, and their own blemished fruits and vegetables, like winter squash, beets, and apples.

"The barns and cold frames keep them warm and sheltered, and they keep my farm from becoming an unplantable mud pit. I have to give the soil a break after they root around in it and fertilize it anyway. I can't just plant vegetables right away."

From May through September, the pigs were free to roam a larger fenced-off area of pasture, hunting for worms, grubs, and other bugs, as well as edible leaves, roots, and grasses—although Chris also fed the pigs grain. In the process, the pigs got plenty of exercise, and in return, their hooves worked their manure into the soil, infusing it with nutrients that would feed his vegetable crops at a later time. In many ways, this seemed like a more appealing model to me than that of the Chapolards, who kept their pigs in barns year-round.

Chris told us he typically slaughtered his pigs at a hanging weight of around 240 to 290 pounds and that it took the animals about eight months to get there. Most pigs I'd worked with so far in Oregon were slaughtered at around 150 to 200 pounds and six months of age, but these were often kept in barns year-round.

I told him that the Chapolards grew their pigs out to about 400 pounds, slaughtering them at around twelve months.

"How can they afford that?" Chris asked, a response common to many farmers I told this to.

"They price it accordingly," Kate said.

"Is anyone willing to pay that price?" Chris asked.

"They sell out every week," I said.

Chris shook his head in bewilderment. "I don't know if anyone would pay that kind of price here."

Dominique asked Chris whether he did all of the farming himself.

"For the most part, it's just me and my wife," Chris said.

Dominique shook his head and said something to Kate.

"He's asking if you had to buy all your land and equipment by yourself, too."

Chris nodded his head yes. "It's very expensive to start a farm," he said.

"Dominique likes to say that if you work alone, you die," Kate said to Chris. "Not only does Dominique work with his brothers and their wives to make their farm run; they are part of a cooperative of farmers who share equipment and a slaughterhouse." Kate briefly explained the GAEC and CUMA models to Chris. "Dominique wants to know why you don't do that here."

"Because in America, small farmers are competitive. We don't want to share our secrets. We don't trust other people. It's every man for himself out here," Chris said, laughing but meaning everything he said.

It was true. Every farmer I met who was raising that 1 percent of animals the non-factory-farmed way was going it alone, buying the same tractor that the farmer next door bought, paying for the same services of a slaughterhouse or butcher like every other farmer, but having little say in how well those services were performed. In America, I was learning, small farmers like Chris worked alone, and they had little control over the process of production once their pigs reached slaughter age.

Dominique shook his head as if to say, *What a pity*.

Kate told Chris she'd send him Dominique's ham recipe and some information about how the GAECs and CUMAs worked in France, but both Chris and I said we thought such models would never fly in America.

"But maybe you're right," Chris said. "Maybe I need to start talking more to my neighbors."

FOR THE WORKSHOP the next day, Dominique butchered one side of pig for the audience, and then Adam demonstrated his very different approach on another side. Michael Ruhlman, Kate, and I provided commentary, comparing and contrasting the two styles. A new charcuterie business in town, Olympia Provisions, the very first USDA-inspected salumeria in Oregon, passed slices of their charcuterie out to the audience of about forty people while they watched Dominique and Adam work. As soon as Dominique and Adam finished, the audience began peppering them with questions.

"Why can't I find coppa in my grocery store?"

"What do I do with trotters?"

"Nitrates. Are you for or against them?" Michael and Adam were for them. "Food safety is paramount," Michael said.

Dominique was against them.

Jo and I snuck a few pieces of Dominique's *saucisson* onto the charcuterie plates we were passing out to the audience and asked them to compare. Elias Cairo, one of the owners of Olympia Provisions, and

Dominique engaged in friendly banter about the differences for the audience, contrasting age, feed, and breed.

Michael wrote about his experience later and managed to capture what I think it's safe to say all of us in the room felt.

"Three hours of intense interaction with people who truly care about this world, the earth and the animals, who care about cooking, about serving people, who do it the hard way, the long way, these grounded wonderful, big big souls. When I walked out of there, I felt as if I'd come out of a world that was impossibly good, could-never-happen good. . . . Swear to god, I wanted to collapse right there at NW 8th and Burnside and weep."

Despite the challenges of the past year, I'd created an experience of impossible goodness with the help of my growing family of mentors—wonderful, big souls who not only wanted to do it the hard way but saw this as the only way. It didn't occur to any of us to try and find any other way around it.

THIRTY-THREE

In the early months of getting to know Jo, she'd often talked with great reverence about a man named Bob Dickson. Bob, she told me, had headed up the meat science program down at Oregon State University, in Corvallis, for twenty-five years, then consulted with meat companies all around the world before moving on to head up a local USDA-inspected, Animal Welfare Approved slaughterhouse and meat-processing facility outside of Portland.

A lot of the meat that Jo received at the Pastaworks meat counter went through Bob's facility, so she'd somehow charmed her way in and met Bob, who'd happily shown her around. At my request, Jo arranged for me to meet him and take a tour of the facility. And so, with a few Portland Meat Collective classes under my belt, and Kate and Dominique having come and gone, I finally entered my first American slaughterhouse.

A pastoral white wooden fence bordered the perimeter of the complex of buildings that housed the facility. Beyond the fence lay fields of grass, the sort of fields that looked as though they should have animals grazing on them, though they did not. From the outside, it resembled the abattoir in Gascony, albeit about five times larger. Jo told me before we'd arrived that this slaughterhouse was considered one of the larger of the seven or so USDA-inspected slaughterhouses we had in the state at the time, even if, compared with slaughterhouses owned by any of the major meat conglomerates around the country, it was actually quite small.

From the parking lot we followed the signs that said OFFICE and walked up a set of narrow stairs lit by a lone fluorescent bulb.

An older woman with bleach-blond straight hair sat at a desk stacked with baskets of invoices, a very ancient-looking computer, a phone with dozens of blinking lights, and a sign that read TALK TO THE HAND.

"Hi, Debbie. How you been doin'?" Jo said in her disarming, golly-gee Southern accent, as if she and Debbie had been friends for ages. "We're here to see Bob. This is my friend Camas."

I extended my hand and smiled at her. The woman stared at me suspiciously.

"So nice to meet you," I said. She gave me a very light handshake without saying anything, got up from her chair, and disappeared down a dark hallway, her footsteps echoing on the linoleum floor.

A couple of Latino guys sat eating their lunches in a makeshift break area off to the side of Debbie's desk.

Jo waved at them. *"Hola,"* she said. I smiled at them.

They nodded at us in unison, then commenced to staring at the empty spaces in front of them, eating in silence, chewing quickly.

Debbie came back, followed by a tall man in his early sixties, with wide shoulders and smile lines creasing his weathered face. He said a friendly hello to the guys eating their lunch, addressing each by his first name, then walked over to us, grinning.

"I've been waiting to meet you," he said to me. "Jo's told me so much about you."

"I've heard a lot about you, too," I said.

We followed him down the dark hallway and into his small private office, which had white walls and the sort of drop ceiling I remember from the classrooms in my high school—which, rumor had it, was originally designed as a prison. We all sat in chairs two feet away from one another, as if huddling over a nonexistent fire to keep warm.

"So," Bob said, "tell me about this Portland Meat Collective."

I told him about the Chapolards, about my belief in transparency as a means of changing how and why we bought, cooked, and ate meat. About my philosophy of whole-animal butchery, which made him chuckle and say, "The whole animal is always butchered, and, trust me, those big meat companies manage to use every part."

"But it's not all used for food," I countered. "Plus, most consumers don't even think of the whole animal as edible, so we sure have to produce a lot of pork chops and tenderloins just to make them happy, and so far as I know, no one has invented a pig made up entirely of loin and tenderloin."

Still, he pointed out, it wasn't as if we just threw away the rest. Between pet food, glue, bullets, and even cigarette filters, every part of the animal went to good use, he said. "The industrial model of meat production is actually incredibly efficient," he said. But he agreed with me that consumers could be better educated about how to eat the whole animal, and that butchers and small farmers working outside of the industrial model might be able to make an actual living if consumers supported them in that way.

I told him about the format for our classes.

"You're doing what I wanted to do twenty years ago but never could," he said. "People just weren't as interested as they are now. I also believe if every slaughterhouse and farm and butcher shop were made of glass, we'd have a very different system of meat production."

I asked him how his processing facility differed from others.

"Well, for one, we're letting you and Jo watch the slaughter. Not every slaughterhouse allows that. We're also trying to work with a diversity of farms, from small to medium to large. And we're trying to satisfy a lot of different needs, from grocery chains to individual consumers. And we are certified as a humane operation.

"Back in the day, slaughterhouses were not nice places to be," Bob said, adding that maybe "nice" is never the right word to use to describe slaughterhouses. "Standards for 'humane treatment' didn't even exist. That *phrase* didn't even exist. Now, thanks to people like Temple Grandin, slaughterhouses, believe it or not, are greatly improved. They are much better."

I wondered what exactly Bob meant. I'd read *The Jungle* by Upton Sinclair long ago and knew that the federal Meat Inspection Act of 1906, which was passed, in part, because of that book, had set basic sanitary, labeling, and safety standards that hadn't previously existed. But did Bob mean things had gotten better since *The Jungle*? Or did he mean that things had gotten better than the horrific—and totally contemporary—slaughterhouse scenes I had seen in undercover YouTube videos floating around online?

"What exactly do you mean by 'better'?" I asked Bob.

"Why don't we take a tour," he said, "so you can see for yourself."

WE FOLLOWED BOB outside.

"First I'll show you what it's like from the animal's perspective."

He walked us over to the holding pens, where the animals settled in the night before slaughter—basically a barn. Since it was already lunchtime, Bob explained, there weren't many animals left—just a few head of cattle, as far as I could see.

A man stood at the entrance to the barn.

"Now watch closely. He's going to open that door and calmly lead one of the steers into this chute." Bob pointed to a long, meandering pathway that led from the barn to the kill room, a few hundred feet away. The pathway was bordered on both sides by tall, gently curving walls made of concrete.

"We used concrete because it doesn't reflect light, which can scare the animals," Bob said. "And we curved the chute in such a way that the animals wouldn't become alarmed by sharp corners, and could not see too far ahead or behind them. The whole goal here is to keep the animal as calm as possible."

I stood up on tiptoe to peer over the curved wall and saw the steer slowly make its way, but Bob motioned for me to crouch down.

"Any surprise in the animal's line of sight might scare him. We want to stay out of sight."

Bob motioned for us to step away from the chute, and we quietly followed him around to the other side of the building, where we could stand without being seen by the animal and watch the kill.

A man on a raised platform stood at the end of the chute, which very gradually ramped up into an open doorway leading to the indoor kill floor. The gradual rise, Bob explained in a whisper, was also meant to prevent undue stress.

From where I stood, I could no longer see the steer. All I could see was the man on the platform. In his hands he held an oblong black metal instrument that looked almost like a short telescope.

"That's the captive bolt gun he'll use to stun the steer before bleeding him," Bob explained.

My heart began to race, and my palms were sweaty. I looked at Jo briefly, but she stared at the man with the bolt gun and did not look back at me.

Bob explained that the steer now stood in a very tight space, with walls right up against his body on all sides, another way to keep the animal calm—*Kind of like a hug*, I thought, aware that I was anthropomorphizing. It was not really at all like a hug.

Considering what was about to happen, everything seemed so very calm and quiet. Out of the corner of my eye, I could sense a lot of activity just inside, but from the outside I couldn't hear any of it. The man calmly raised the bolt gun up toward the steer's head, in a slow, considered manner. I watched the man's eyes watch the steer's eyes. I heard a brief, forceful pneumatic puff of air from the gun, and the steer's head vanished.

Bob told us to look through the slats in the gate in front of us. I could see onto the kill floor. The animal had fallen to the ground, and a man had already stuck a knife into the steer to begin the bleeding process.

At least this is how I remember it. The thing about watching something like this without taking notes—I wasn't there as a journalist after all, but just as a curious consumer—is that all you have to go on afterward is a loosely knotted string of disparate, exaggerated sights, sounds, and smells that dig a zigzagging rut in your memory and lodge themselves under an ever collecting array of detritus. I remember how, after that puff of air from the bolt gun, the man stared down toward the steer from his platform, presumably to make sure he'd done his job correctly, and then, how he turned slowly away and looked out into the empty space in front of him, and how his chest rose and fell with the next breath he took.

I remember thinking I would never be able to find words for any of this. Not because it was bad or good, or better or worse, but because that's what witnessing this particular kind of death does: it brings to light our lack of vocabulary for it. Dario Cecchini, the famous Italian butcher whom Bill Buford wrote about in his book *Heat,* referred to humane slaughter as "a good death," one of four things he felt an animal that is killed for dinner is owed—a good life, a good death, a good butcher, and a good cook. "A good death" seemed appropriate enough, but even this did not quite capture the black hole one fell into after witnessing such an event.

I remember the USDA inspector standing at a table, slicing into a liver with a knife to inspect it for signs of illness. I remember watching a short man strip the hide from a steer hung from the ceiling in a matter of seconds. I remember the cleanness of the space. The steer's massive carcass. The silence of the workers' movements. Their serious faces. I remember understanding then what Bob had meant by "better." This place appeared organized, calm, efficient, and—in every way that it could be—humane. The curved walls. The dull, unreflective concrete. The slow, considered movement of the man with the gun.

I HAVE SAID that slaughter is hard to watch. *Hard* always seems like the wrong word, but I never want to give the impression that it is easy. Without fail, someone always asks me, "If it's so hard, why do it?" I believe that doing something hard makes me a better person, a more realistic and responsible one. That *easy* makes us supporters of a system of meat production gone totally awry. That is what I think Levi had meant when he told me that the day slaughtering a pig no longer felt "a little horrific" would be the day he would cease to eat pigs.

We toured the rest of the facility with Bob. The room where they air-chill their chickens. The room where the beef carcasses age. The room where the sides of pork hang. The room where more men and women—mostly Latino—clad in chain mail and protective gloves, stand elbow to elbow, all day long, cutting pork and beef primals.

Before leaving, I asked Bob if he would teach a class for us.

"I want to do the class here, in this facility," I said. "I want your employees to be there, too, to teach us what they know." I wanted my students to meet the people willing to do the job that they were unwilling to do, the job that allowed them to believe they were exempt from the entire process.

Bob looked at me as we stepped out into the summer light. "There's a few people here who might not see the point in that, but I do. I'll see what I can do," he said.

THIRTY-FOUR

Twelve multicolored roosters were sweating, shitting, and scratching at one another in the back of my car, and I was not headed to a cockfight.

Earlier that morning, on the advice of a farm foreman whom Jo and I had met down in Scio, I'd driven about forty-five minutes south on I-5 to a livestock auction in Woodburn, in search of old roosters, the kind of birds traditionally destined, after a long, healthy life, to be placed in a heavy-bottomed pot, along with button mushrooms and sweet pearl onions, bright-orange carrots, bay leaves, celery, and thyme, and then covered with rich, tannic French red wine. I was looking for roosters so old and tough they needed at least a day of marinating, and another full day, if not two, of stewing. Muscled, well-exercised guys who'd developed their hefty, toned stature not by way of steroids or limited exercise in cramped indoor spaces, but simply by living how a bird should live—outdoors, moving, jumping, flapping, pecking, shitting, eating, fucking.

I'd recently learned from a chicken farmer that chickens raised for meat in America are typically slaughtered at around five to seven weeks old. I was looking for five to seven *years* old. But when I'd bid on these twelve birds, their formative years were largely a mystery to me. I'd had to look these birds up and down and rely on intuition.

What I did know for certain was that these twelve roosters were destined for twelve different pots that would be stirred by twelve different people, the kind of people who wanted to learn how to kill their own dinner. I'd dubbed the class, my first chicken slaughter class, "Real Coq au Vin," after the classic wine-infused French stew, a recipe that traditionally calls for older roosters. By "real" I also meant that everyone in the class would start from the beginning—with a live bird. Levi Cole and I would teach the students how to kill the birds as painlessly as possible and then how to turn the birds into food.

I'd split the roosters up evenly between two large dog carriers I borrowed from Levi and a few cardboard boxes that I'd carved breathing holes into. It was no verdant pasture, but it was at least triple the amount of space provided to the chickens I'd seen riding in open-air cages stacked twenty deep on the backs of many a large semi driving down I-5, most likely factory-farmed birds headed to slaughter. Nevertheless, I was worried about the birds. I worried that they were thirsty. I worried that *they* were worried. I also worried that teaching twelve students how to kill these birds was a really bad idea. What if someone didn't kill their bird right? What if people cried?

What if? What if people did cry? Why this inclination to ensure that everyone was comfortable, that no one had to feel or think too hard about anything? Wasn't the point of this class precisely to help people understand that things *can* go wrong, that killing animals well required skill, practice, and commitment, and that we might be better off if we acknowledged that fact?

After fifteen minutes of driving alone with the roosters, my friend Jill drove up behind me, honking and flashing her lights before pulling into the lane next to me. After being let go from the city magazine and then giving birth to her daughter, Jill had enrolled in a film course, and for her final project she was making a short documentary about my class. Her classmate Amanda rolled down Jill's backseat window and pointed a video camera at me.

I flashed Amanda an awkward smile, the kind of smile that says, *I know that you know that I am unsure of what I am doing here, but I'm going to go ahead and smile anyway,* and kept driving as Jill matched my speed.

EARLIER THAT MORNING, after we'd arrived at the auction's vast dirt parking lot, filled with gooseneck trailers and pickups, semis and four-wheel drives, I'd wondered whether bringing a film crew was a good idea. And when the gruff, permed woman working the main office, clearly miffed by our presence, chewing her gum methodically, informed us that the auction had a strict no-camera policy, I was secretly relieved. But then Jill intervened with her Texan brand of golly-gee charm and changed the woman's mind.

"This is just for a class," Jill said innocently. "We're students. We're just practicing! But we've got consent forms and everything, just like the real thing!"

The woman sighed dramatically. "Fine. Go ahead," she said between chews. "If you're gonna bid on birds . . . [*chew, chew, chew*] . . . you're gonna . . . [*chew, chew, chew*] need one of these." She handed me an index card with a number on it.

And then we entered the pungent and labyrinthine complex of open-air barns, fenced outbuildings, and cordoned-off patches of cracked asphalt and dirt that made up the Woodburn Livestock Exchange. We stopped first inside a covered, rodeo-like arena where a man paraded a hulking steer in front of a crowd of thirty or so men standing in dirty jeans and rubber boots, cowboy hats and Carhartt vests. As we watched the proceedings, thirty pairs of eyes darted from the steer to Jill, Amanda, and me, the only women in the room.

Next door, more men picked through row after row of small farm equipment laid on the dirt ground in flea market fashion. There were big silver shears, lengthy coils of green hose, rolls of used chicken wire, banged-up aluminum troughs, disassembled rabbit hutches, half-used bags of feed, and balls of frayed twine.

Outside, in a dry field cordoned off by temporary plastic fencing, men stood in groups of two and three admiring giant, used blue and green tractors.

While the sounds of crowing roosters and mooing cattle bombarded us, the rhythmic chants of the auctioneers, calling out from their respective jurisdictions, hypnotized us into a quiet lull. I had no idea what they were saying. Maybe I had no business bidding on twelve roosters with my students' money. But would the students have otherwise done this on their own? Someone had to do it.

"Excuse me," I said to an older Vietnamese woman, the only woman I'd seen since we left the auction office. "Where are the birds?" Jill pointed the boom mike between us.

The woman shrugged, not understanding me.

I flapped my arms like a bird. "Chickens? Roosters?"

She laughed and pointed behind her.

We rounded the corner of the main barn and came upon a small, open-air shed stacked floor to ceiling with cages of birds. Each of the five dozen or so roomy wire cages temporarily housed one, two, or three birds. A trio of fluffy white chickens with decorative crowns of feathers huddled together in one cage. A classic black-and-white hen clucked down at them. Several night-black cocks nearly three feet tall crowed to the crowd. At least I thought they were cocks: they had the red cockscomb I'd always associated with a rooster's essential identity, but then I noticed eggs in some of their cages and I began to seriously doubt my assessment criteria.

The people around me talked to one another, pointing at various birds. I assumed they knew how to tell a rooster from a hen better than I did. In the other areas we'd wandered through, I'd seen mostly white and Latino men. In the bird room, I was surrounded by men *and* women, speaking to one another in their native languages of Russian, Spanish, Vietnamese, and Chinese.

A red-faced white man walked in, his plaid flannel shirt tucked neatly into a pair of pleated khakis that clung tightly to his bulbous paunch, and the room shifted noticeably. We leaned forward as a group, nearly one body with many arms and legs, angling for the best birds, the plumpest birds, the oldest or the youngest or the prettiest.

I had my eyes on that trio of fluffy white ones. They weren't tall, but they were wide in girth, so I assumed they'd be meaty. I had no idea how to tell their age, but I figured the auctioneer would tell us. He didn't. At the very least, I was sure they were not Cornish Crosses. I'd recently learned that most birds we eat in America are of the Cornish Cross variety, a crossbreed built to grow fast—and develop large breast muscles in a small amount of time—and thus be ready for slaughter at that five- to seven-week mark. Several farmers who'd attempted to pasture Cornish Crosses had told me that if the birds were allowed to live past that age and their breeding lines hadn't been closely monitored, their small legs, unable to hold up their heavy upper body, often gave out. Or they'd have heart attacks. Or they'd die of thirst because they couldn't walk to water. I wasn't interested in Cornish Crosses. I wanted a sturdier bird.

The auctioneer opened his mouth and began speaking nonsense.

"Numma-numma-NUM-NUM-bwana-bwana-DOO-DOO-canna-canna-FIE-FIE."

I thought I saw the auctioneer point at my white fluffies, heard the number ten, panicked, and foisted my number up in the air. I felt like I was back in France again, making wild and incorrect assumptions about the meaning of what I'd heard.

More jumbled sounds burst forth from the auctioneer's mouth. I raised my hand a few more times and then "SOLD!" Jill let out a whoop. I'd forgotten, momentarily, that I was being filmed. The crowd's collective body, with its one hundred eyes, turned to blink at me, then at the cameras and microphone, then back to the auctioneer. Had I just bought a bird? I wasn't entirely sure.

A young woman separated from the group and stepped toward me. She was about my age, maybe a little younger, in jeans, a faded jean jacket, and muck boots.

"You just bought those three birds for twenty dollars each. You know that, right?"

Her lips were frosted with a gloss whose pink hue reminded me of the colors of the Lane County Fair, which I'd grown up going to in the early eighties. Like sunlight shining on cotton candy. Light. *Please be my light,* I thought.

"I have no idea what I'm doing."

"Twenty dollars is way too much for those birds."

"Which birds did I buy?"

She pointed to the fluffy, thick white birds I'd been eyeing.

"I wanted them! Perfect!"

"What do you want *them* for?"

"Umm, to eat."

"Those are Silkies. Nothing but feathers. You won't find any meat on those bones."

"What? Jesus." I was already halfway through my budget and I'd bought only a quarter of what I needed. "Tell me how this works. I'm teaching a slaughter class and I want old roosters. How do I know for sure they are roosters?"

She looked at me funny, held her hand up to silence me, watched the auctioneer's mouth for a few seconds, and then surveyed the crowd with her blue eyes. The man standing in front of me held his card up.

"Put your number up," she told me. "Starting bid was five dollars."

I raised my hand. The auctioneer looked straight at me and pointed.

"*Siiiix. Siiiiiiiix. Numma-numma. Bwana-bwana. Siiiix.*" Another hand went up.

"Bids are going up by twenty-five cents now. Put your number up again. This is a big rooster," she said, pointing toward a yellow-and-black fella.

"*Wuuuun. Twiiiiigh. Sold!*" The crowd turned and blinked at me once again. My new friend explained to me that I'd just bought the bird for six dollars and fifty cents.

"This is way too fast," I said to my new teacher. "Thank you."

"So are you on TV or something?" she asked.

"Um. No. My friend is just filming for her film class."

"You guys from Portland, then?"

"Is it that obvious?"

She didn't answer. "So you're teaching city folk how to kill roosters?"

"Yes. That's the idea."

"Truth is, we could use the same classes out where I live," she said.

After she helped me garner a few more winning bids, I'd met my quota.

"Now what?" I asked her.

"Get your total from the guy and go pay at the office. Then you get your birds."

I hadn't really thought about this part. Now I was going to have to transfer, with my own two hands, nine very large, feisty-looking roosters and three not-quite-so-menacing white ones into my car.

"Right. Okay. I'll be right back."

"You're gonna need my help," she said.

"I probably am."

WHEN I RETURNED, my new friend spoke in hand gestures to two older Chinese women who looked to be in their eighties. About half of the crowd had left. Jill and Amanda rolled camera.

I didn't want to bother my new mentor any more than I already had, so I brought one of the dog carriers over to my twenty-dollar Silkies. When I unhinged the door latch, they backed away into the corner. My palms were sweating. This was not something I did every day, wrestling chickens into the backseat of my car.

As I reached into the cage, the sharp wire edge of the door caught my forearm and traced a long bloody line from my wrist to my elbow.

"Shit," I muttered.

I tucked my head, neck, and shoulders into the cage and got hold of one, albeit precariously. But before we both made it out, the bird grabbed on to the mesh floor with his claws. His body was quite tiny—there would indeed be very little meat on these bones—but his strength greatly outweighed his stature. Jill and her crew moved closer to me with their cameras. The cut on my arm began to burn. My face burned, too.

I pushed the bird's body in the direction of his grip, to confuse him into thinking I was letting him go, and this caused the Silkie to briefly loosen his claws such that I was able to fully extract him. I tried to wedge open the door to the dog carrier with my foot, but the bird was struggling out of my sweaty hands, a fan of white feathers thudding against my chest.

The young woman returned at exactly that moment and opened the door for me, closing it swiftly after I'd gotten the bird safely inside.

"You have to do it fast," she said. "Don't loiter in front of the cage. Don't think too hard. Just approach, open it quietly, move your hands up real fast so they don't know what's happening, and then grab them around their wings so they can't flap."

"Okay. Got it," I said. Blood dripped from my arm to the ground.

"And watch their claws. Those'll get you worse than that cage got you," she said.

And all this before you've killed the rooster for dinner.

THIRTY-FIVE

After they'd spent a few nights ensconced in a makeshift chicken-wire run that Andrew helped me build in his side yard, I transferred the birds, again somewhat awkwardly, back into their carriers and drove them to the same urban farm where Jo and Adam and I held our first pig butchery class.

By the time I arrived, Levi already had two large pots of water going over a couple of propane burners he'd set up on the paved driveway near the barn.

"Good morning!" Levi said. He had a habit of resting his hand on his throat whenever he was nervous, and he was doing it now.

"You ready to kill some chickens?"

Was anyone ever really ready?

Together Levi and I walked the carriers over to a tall patch of grass that a few of the farm's employees had cordoned off with temporary fencing. One by one, the roosters strutted out into the makeshift pasture and began to run in circles and figure eights, exploring their new landscape. They also occasionally started fighting—whether playfully or seriously, it was not entirely clear.

Jill arrived with her camera. I still felt conflicted about all of this being filmed, especially the slaughter part. Would the presence of a camera somehow abstract the whole thing for the students, not to mention anyone who would end up viewing Jill's movie?

Levi found a straggler left in one of the carriers, which he tilted in order to coax the bird out. The bird slid to the ground but did not stand up. His breathing seemed labored, his feathers askance, wet almost. Something was wrong. I became aware of how hot it was outside—it was June, not the best month for a chicken slaughter class. Sweat began to drip from my armpits. What had I done?

Jill moved in closer with her camera.

"We're going to have to kill this one now," he said.

"What's wrong with him?" I asked.

"It looks like maybe he broke his leg. Might have overheated, too. Not sure," Levi said. The bird wasn't dead, but he looked like he might be better off if he were.

Maybe I'd left them in the car too long. Maybe I'd put too many in one carrier. All of my worrying about the birds had not prevented this casualty.

Levi cradled the bird in his arms and pet his head.

"What happened to you, old fella?" Jill zoomed in on Levi. "Let's not film this one," he said. Jill didn't turn her camera off.

I followed Levi to his table of tools: scalpels, blunt-nosed scissors, latex gloves—all instruments he used to keep people alive in his day job as a nurse. Feeling guilty and mournful, I held the bird's body upside down in my hands, as Levi had taught me. He killed the bird quickly. I could feel the slowing rhythm of a body going from tense to slack, from live to dead in less than thirty seconds.

"Well, that sucked," Levi said. He was right. It *was* difficult. *Of course it was*. But it would have been worse for the bird had we not killed it right away.

AFTER A FEW MORE PREPARATIONS—laying out buckets for guts and feathers, setting the farmhouse table, making sure we had enough of Jill's media releases—the students began arriving. A local, and very opinionated, food blogger showed up with his own video camera, followed by a man who immediately began quizzing me about my knowledge of a classic coq au vin recipe penned by the famous French chef Paul Bocuse. I wasn't familiar with Bocuse's recipe, and the man seemed disappointed. I got the sense that he was convinced that the class would not be authentic enough for his taste, but I appreciated his clear knowledge of French food. He'd even brought a platter of thinly sliced head cheese that he'd made himself.

A few students introduced themselves as farmers—some rural, some urban—and said that, while they'd started raising chickens, they didn't know how to properly kill them when the time came. A timid stay-at-home mom said her kids were entering college and she was interested in getting into the meat business. At the last minute, a young woman confined to a wheelchair was loaded out of the back of a van by another woman—her "surrogate"—whose job would be to convey, as best she could, to the young wheelchair-bound woman what it was like to slaughter a bird.

And then we began. Demonstration first. Scalpel. Palate. Carotid arteries. Check the eyes for activity. Levi spoke calmly, using medical terms to explain what was happening to the bird's body. As the bird bled out, the students remained silent, but interested, alert, curious. Our shared witnessing brought us all together. It also forced each of us to separately confront our own singular fragility and mortality in the face of another animal's death.

Levi explained that this was the most humane way to kill the bird and that by humanely dispatching the animal, we were ensuring that it did not unnecessarily suffer, but we were also ensuring that the meat of the bird had the proper amount of lactic acid, thereby preventing unappealing flavors and textures.

"Although," I added, "some of these are older birds—at least I think they are—so even if you kill them well, they are going to have a good amount of bite if you don't cook them right."

The bird was a carcass now, in technical terms, and not so much a body, and so Levi moved on to the next step, showing the students how to dip the carcass into a pot of nearly boiling water and swirl it around gently to loosen the feathers.

"Make sure you know the bird is dead before you do this," he said. "If he's not, he'll feel the hot water and start flapping his wings, and then you'll feel the hot water, too."

Levi walked us through feather plucking, which transformed the bird into a pale, naked, vulnerable form.

"Why don't you all partner up and get this far, and then I'll show you how to eviscerate them," Levi said.

"One last thing," I said. "You need to either commit to doing this or don't do it at all. Hesitating means the bird will suffer. If you are doing it wrong, we will intervene. If you can't do it, we'll do it for you, or, if you like, you can take your bird home still alive. It's your bird now." I sounded cold, but I needed them to feel the weight of the responsibility, even though the burden of these twelve birds' deaths, in many ways, fell to Levi and me.

AND THEN, ONE BY ONE, we watched twelve adults awkwardly attempt to pick up their especially agile, wily birds. With their legs bent, their backs hunched, and their arms and hands fully extended out in front of them like Frankenstein, they'd walk slowly toward a bird, trying not to scare it, and then lunge at the last moment. It mostly seemed to work.

When half of the students had gotten hold of their birds, we instructed them to partner up with another student and begin the slaughter process.

They all stood motionless, blinking at us, holding their birds close to their chests, caressing them. It's one thing to watch a slaughter; it's another thing to do it yourself. Most of them had never had to do this before. I'm pretty sure all of them wondered if they were the right person for the job. I wondered, too.

Levi and I moved from student to student, helping them work through the process. Their confusion, their frustration, their worry that they were doing it wrong played out in mostly identical ways.

"I don't *see* the artery. Where is it?"

"You won't necessarily be able to see it," we said, "but if you push the feathers up like this and put your finger to its neck and feel its pulse, that's where the knife should go."

"Am I hurting the bird?" To which we responded, gently, "You'll know if you are hurting it," or, "It's probably uncomfortable, but you are not hurting it." Or, in a few cases, "You will hurt it if you don't finish this now. Stick the knife in fully. Now. Do it," we would command.

A few hands shook. A few faces turned red. One young woman said she didn't think she could do it and paced back and forth for a while, but then she did it. After her bird bled out, she set it down on the ground and walked away from the group. I followed her.

"Are you okay?" I asked her.

Her eyes appeared watery.

"I'm okay," she said. "I just feel . . . different . . . somehow."

All of the students were quiet, somber, respectful.

They dipped their birds in the hot water and then sat on overturned buckets and stumps, plucking their birds, talking about their families, their jobs, how their grandparents and great-grandparents once lived, and, finally, about food. I apologized to the students who'd ended up with the Silkies—they truly were scrawny—telling them the story of my travails at the livestock auction.

"Hopefully they will be so full of flavor, it won't matter how little meat is on their bones," I said.

In the film that Jill made, we then see the students gathered around Levi as he cuts an incision between the bird's legs, puts his hand inside the bird, pulls out a fist-size mass of brightly colored organs, and shows the students the intestines, the spleen, the liver, the kidneys, the gizzard, the heart. While he's talking, the camera pans to the woman who hadn't been sure she could kill her bird. She looks to the faces of each of her classmates for confirmation of her own clearly complex feelings. Her lips and nose are drawn slightly upward, as if in disgust.

The camera cuts to the hands of the students as they hesitantly cleave heads and feet off and then pick through their birds' organs, marveling at the surprising colors and forms. The surrogate for the woman in the wheelchair cups her hand underneath the wheelchair-bound woman's gnarled hands and places a chicken heart in them, then the liver, then the lungs.

At one point we discovered that some of our roosters were in fact not roosters at all. The Francophile found a beautiful, fine string of tiny yellow egg yolks in his bird, each at a different stage of development.

"In France," the Francophile said, "these are considered delicacies." Then he picked one of the tiny yolks up, tipped his head back, and put it in his mouth. A few people laughed uncomfortably. Others began asking questions. "Is that really how they eat them? They don't cook them first? What does it taste like?"

This prompted Levi to remind everyone that salmonella (as well as campylobacter) can occur naturally in chickens' guts—not to mention chickens' feet, feathers, and beaks—and that, while not all strains of salmonella cause us harm, it was important to take precautions by keeping the guts away from the edible parts of the bird's carcass, keeping the carcass as cold as possible during and after processing, and cooking chickens to the right temperature.

As we walked up to the farm's main house, each student carrying a plastic bag of freshly killed chicken, the students continued to ask questions.

"So how do you cook a liver?"

"Is the heart good to eat?"

"When are you going to teach a pig slaughter class?"

These classes were taking the students down the same rabbit hole that my time in France had sent me down. None of us had even known how many questions we had—or which questions to ask—until we'd taken this one step into the real-world problem of how to kill an animal for dinner. These were questions whose answers, not that long ago, were all around us, passed on from grandparent to parent to child. But who was there to answer these questions now? Purdue? Tyson? They weren't answering their phones. And I could only assume the way they raised and killed and processed their birds was a far cry from how Grandma did it.

In the farmhouse, we poured wine into glasses and spooned the coq au vin I'd made into bowls. We sliced crusty baguettes and passed around ramekins of Levi's chicken liver mousse and the Francophile's buttery head cheese. As the students ate, I talked about using old birds for the dish, how, even though I'd tried my best to buy what looked like older birds at the auction, I could not be sure of their age.

One student asked whether there were farmers who specialized in older birds. I said that the demand for young, tender, pale, mild chicken was too high in America. "If you guys want older birds, you'll either have to produce them yourself or start demanding them from farmers you know. But you'll also have to pay a higher price."

The meal wasn't perfect. The meat was a bit dry and it flaked into minuscule strands, as opposed to adhering to the bone of each cut, most likely because the bird I'd used had not been as old as I'd wanted to believe it was and I'd overcooked it. The Francophile made sure to complain once again that I hadn't used a more authentic recipe. But we'd been as authentic as we could have been, given our limited resources and the gap of lost knowledge we were trying to bridge, given the fact that we lived in America, land of bland, boneless, skinless chicken breasts.

AFTER THE CLASS, I searched online for Paul Bocuse's coq au vin recipe. The recipe's headnote read:

"He was a tough guy and he therefore required slow cooking. I must say that in yesteryears in Burgundy, roosters were at least three kilos and had run a good deal before making their way into the pot. Today roosters found in the market under the name of coq au vin are not as muscular and probably require shorter cooking times. But, either way, the wine needed to prepare this dish should always, obviously, be from Burgundy."

Even Bocuse could no longer find the right bird for the job.

At least I had used a good Burgundy for the dish.

A month or so later, Jill and her fellow students debuted their final projects for a public audience at one of the many old movie houses in Portland that serve beer and pizza. Andrew and I showed up just as Jill's film, which she'd titled *Good Bird*, began. In the beginning, I am crouched underneath the low chicken-wire roof of the chicken run in Andrew's side yard, unable to stand up straight, attempting to corral the roosters into their respective carriers. I keep snagging my bright-green T-shirt on the chicken-wire roof and yelping, a large creature stuck in a shelter that is way too small. I was reminded of the scene in *Alice in Wonderland* in which she drinks a magic elixir that makes her grow to the size of a giant, while standing inside a very small house. My body in that chicken run, getting caught on the chicken wire, reminded me of the awkwardness of my project, its vulnerable, maybe even impossible, nature. I scrunched down in my seat.

By the time that sick, suffering bird made its appearance on the big screen, I had already left the theater and hidden in the cramped women's bathroom. I couldn't watch the rest of it. It had been a respectful film, but not a serious one. The charming, madcap lightness of the scene cuts, the timing of the whimsical accordion music Jill had chosen, had even made the audience chuckle a few times, and yet the burden of all these birds' deaths, captured on film, overwhelmed me. I didn't want anyone to think I took any of it lightly or had been foolish or even careless in my endeavors.

Andrew came and found me.

"What's wrong? Why are you hiding?"

"It's too public," I said. "I don't want it to be that public."

THIRTY-SIX

By the summer of 2010, a year after I'd returned from France, my Portland Meat Collective classes—from pig and beef butchery to sausage making—continually sold out. I was able to start paying myself. I was also making decent money working at Evoe and doing bookkeeping for my brother, I still wrote occasionally—but only when I wanted to—and I was finally off unemployment. Yet I still felt something was missing. I wanted to work in a butcher shop. It was what friends and family had imagined when I told them I wanted to "be a butcher" and then run away to France. It was what I'd imagined, too.

So when Matt and Stu, the two guys Jo worked with at the Pastaworks meat counter, decided they were ready to move on—or maybe it was decided for them—I applied to work there a few days a week, in addition to everything else I had going on. Jo put in a good word for me, even though it was probably a disastrous idea for her and me to work together more than we already did. We were still struggling to maintain a platonic friendship, occasionally giving in to our more complex feelings for each other, and whenever this happened I would fall into a guilty rage, which I aimed at Jo, as well as at Andrew, without his understanding where it came from.

I started my first shift before Matt and Stu left. Dressed in yet another white butcher's coat two sizes too big for me, I learned one very important piece of information from them: don't talk while operating the meat slicer. Matt and Stu went out of their way to tell me this, but once I started up that spinning metal blade of death myself, Stu made sure to gab at me nonstop, and I promptly sliced the tip of my right ring finger off. I wrapped a foot-long strip of gauze around my gushing wound and topped it off with a latex glove, ready to keep working, but the store's general manager insisted I go home for the day.

On my way out, I glared at Stu, who laughed and yelled, "Don't talk so much next time, honey!"

Thankfully, Jo trained me for my second shift. After a quick tour of the meat case, she quizzed me. "What do you think we sell the most of?"

"Ground beef?" I guessed.

"Close. We sell a shit-ton of sausage," she said, arranging two dozen spicy Italian links in a plastic tray. "Next is ground beef and pork. Then bacon. Then steaks and chops and tenderloin. Then leg and shoulder roasts. Then stew meat. Then charcuterie. Oh, and lots of chicken breast."

I asked Jo where the rest of the parts of the animals went, and she explained that they rarely got whole animals in. Instead they ordered only the subprimals and retail cuts they knew they could sell, like beef and pork tenderloins, whole pork bellies to make their own bacon, plus boneless pork shoulder for grinding or sausage.

"Where do all the parts come from?" I asked.

"Various distributors. A few local farms and ranches, but only those willing to sell us primals or subprimals. Most of the pork comes from Carlton Farms," she said. "But the pork shoulders we use for sausage come from Mile End Farm." Mile End, where we'd gotten pigs for our first Portland Meat Collective classes.

"What's Carlton Farms?" I asked. I'd seen the name on a lot of "farm-to-table" menus around town.

"Most people think it's a farm in Carlton," Jo said, referring to a town about an hour southwest of Portland. "But it's really just a slaughterhouse that buys live animals from other farmers, slaughters and processes them, and then sells the parts wholesale."

"Who are the farmers?"

Jo shrugged. "I keep asking, but they never tell me." Many years later, I had the chance to ask the president of Carlton Farms, and he wrote, in an e-mail, that they buy "natural pigs" (his use of quotation marks) who are raised by "family farmers" (again, his use of quotation marks) and that most of them came from the Hutterites, "who specialize in agriculture as you may know." He did offer that they also worked with Oregon farmers who are raising heritage pigs, but I gathered that these did not make up the majority of their sales. When I asked him how the pigs were raised and which Hutterites he was referring to, since so far as I could tell, Hutterites, an ethnoreligious branch of Anabaptists, lived all over Canada and America, he offered no further details. To this day, well-meaning chefs in Portland put the Carlton Farms name on their menus when they want to tout their "local ingredients." In truth, most of the pigs were likely only *killed* locally.

"All right, honey, we gotta make at least a hundred pounds of sausage today, enough to last us for the week."

Jo walked me through the quirks of their ancient meat grinder. "Don't screw the arm on too tight. . . . If it starts to rattle, knock it on the back side with the flat edge of this cleaver." Then she set me up in the front window with a tall manual sausage stuffer.

"If a pretty woman like you stuffing sausages in the front window doesn't bring in the customers, I don't know what will," she said.

She was joking, of course, but I wondered whether something similar hadn't crossed the minds of Pastaworks' owner too. One of the first customers I helped, a woman, told me how much friendlier the counter seemed with women behind it. I could relate. As a customer on the other side of the meat counter, I'd always found the men who worked there to be gruff and intimidating, even a touch angry.

BUT NO WOMEN applied for Matt's and Stu's positions, and soon Jo and I had some new colleagues to contend with. Our first new recruit, Mike, was in his late twenties, but he'd worked as a butcher for a decade—on kill floors, in processing facilities, and at grocery store meat counters up and down the West Coast. He was a diminutive, sweaty creature with a penchant for death metal and a Napoleon-style chip on his shoulder that was much too large for his tiny frame. Nonetheless, unlike us, he was a real butcher, and we were excited to learn from him.

We convinced Pastaworks' managers to let us get a couple of whole animals for Mike to work on, and for the first few weeks he patiently answered our questions as we watched him break down lambs and pigs. But over time, he began to withhold his knowledge, especially as Jo and I began getting more media attention for the classes we were putting on. When one national food magazine identified Jo and me as "butchers," Mike was quick to tell us, "You're not butchers. You're meat counter workers." He was right, of course: we weren't yet butchers. We were still learning. But in his words I detected a more permanent classification: we were service workers, the female kind, and that's what we would always be.

A few weeks after Mike arrived, Pastaworks brought in Jonah, a born-again Christian with a Hare Krishna haircut, a Charles Ingalls sense of style, and a whole lot of self-conscious Portland hipster thrown in. He'd recently apprenticed at an organic farm in Colorado, where he'd learned to speak the language of "sustainable meat," whole-animal butchery, and charcuterie. I gathered he fancied himself a sensitive man, but when he addressed female customers as "darling" and "sweetie," it sounded mostly lecherous. The term *mansplaining* had not yet reached fever pitch in popular culture, but it goes a long way to describing his behavior toward Jo and me.

I didn't know if Jonah had read *The New York Times* butcher-as-indie-rock-star article, but he sure acted like one. He liked to bring in big hunks of salami and pancetta that he'd cured in his basement and hang them in the window of the butcher shop for all to behold—trophies that apparently attested to his authenticity as a real butcher.

The presence of feisty, vocal women who wanted to be butchers and had strong opinions about how the shop should be run caused Mike and Jonah much consternation. A month or so in, I took it upon myself to create new communication protocols and order sheets. When Jo and I sat down with Mike and Jonah to review the new procedures, they rolled their eyes and slammed out of the meeting. They were having none of it. None of it, that is, except for the clipboard I hung on the wall for us to leave written messages for whoever had the next shift.

One early morning shift, Jo and I encountered this note from Jonah:

"Yo Bitches, I had a rough night so don't get all up in my shit about not washing the dishes or cleaning the counter." The expectation being that we would, of course, just shut our mouths and do the work he'd failed to do without complaining.

Jo and I mostly rolled our eyes at all of this. Unlike these boys, we'd come from professional careers in which we'd quickly learned our way to the top and managed other people. Perhaps naïvely, we assumed we would be able to do the same in the world of butchery. But it became increasingly clear that that wasn't going to happen here. Instead, the meat counter felt more like grade school recess—girls get the swing set, boys lord over the play structure.

STILL, WE DID MANAGE to learn a few important lessons alongside them.

One day, I walked in for my afternoon shift and saw Mike and Jo staring at a boneless pork shoulder on the counter.

"Does this look different to you?" Jo asked me.

The hunk of pork shoulder was extremely pale, and much larger than the ones we normally bought for making sausage. Very little fat ran throughout the knots of shoulder muscles, and a film of viscous, snotty-looking red fluid covered the outside. It didn't look or smell very appetizing, either.

"I think our order got mixed up," Mike said. "This doesn't look like Mile End's pork shoulder."

Jo knew someone at the facility where Mile End's pigs were slaughtered and further processed, so she called them to investigate further.

After hanging up, Jo looked ashen.

"I can't fucking believe this," she said. "They said that the order was right. That the meat looked different because it was IBP meat."

"What's IBP?" I asked.

"Iowa Beef Processors," Mike said. He explained that while IBP was owned by Tyson, one of the four major meat conglomerates in America, Tyson still used the IBP name on its commodity pork and beef.

"It's shitty factory-farmed meat," Jo said. "Apparently Mile End sells IBP pork under their name all the time to supplement when they're running low on their own pigs, which is, apparently, quite often."

"How can that even be legal?" I asked.

I'd recently become friendly with a local USDA inspector and decided to call her up. When I told her what had happened, she laughed at my naïveté.

"It's called co-packing," she said. "And it's perfectly legal. Any farm can buy any USDA-inspected meat and sell it under their name without having to disclose where it came from. They only have to tell you what slaughterhouse it went through, which is indicated by the USDA stamp on the meat."

"So there's no way for me to trace this shoulder back to wherever it was raised?"

"No easy way. Oh, and there's plenty of farmers at the farmers' market selling meat this way."

"That's not considered fraud?"

"It's not. And probably none of their customers would ever be able to tell the difference."

I thought back to what Kate had said about Portland charcuterie. *Not bad, but American.*

This felt so goddamned American. You could call yourself a farm, design yourself a logo with white picket fences, walk around with photos of your happy, humanely raised pigs, and no one would know that you were really selling them factory-farmed commodity pork. No wonder Mile End had been able to sell me pigs at such a low price on such short notice for my first class.

Tricia, from Mile End, denied everything.

AFTER THE PORK SHOULDER debacle, Pastaworks let Jo and me order a whole pig from a new and, we hoped, more trustworthy farm, with the goal of figuring out whether we could sustain a whole-animal butchery program and stop having to buy meat from distributors and co-packers. The pig farmers grew their own grain to feed their pigs, and raised their pigs in open-air barns just like the Chapolards did. This time we knew to ask them if they co-packed, although they weren't legally obliged to tell us the truth.

We made sure to get the two sides delivered on a day when Mike and Jonah weren't there, and together we came up with a fresh-cut list that very much resembled that of the Chapolards. It'd been a while since I butchered a pig, but I'd learned much at the two dozen or so PMC classes I'd held, and the process felt so much more instinctual now. In fact, for the first time, I managed to remove the rib bones and spine cleanly from the loin, so that I had a beautiful, long muscle to work with. I left it whole and proudly displayed it in the case, just as the Chapolards did in theirs, with a sign next to it. WE LL CUSTOM-CUT YOUR CHOPS, it said.

Only one customer asked about it that day.

"Is that tuna?" he asked.

"No, sir. That's a fresh pork loin," I said. "I can cut you off as thick or thin a chop as you like. Or I can cut a nice roast for you and tie it."

"Can you just give me one of those?" He pointed to a tray of slightly graying, but more recognizable, bone-in pork chops we'd cut from the other side of the pig. Something about seeing the whole muscle had turned him off. After a few days with no other inquiries, I cut my beautiful whole loin muscle into chops and they finally sold.

A few customers did buy the pig ears for their dogs. And a few took a chance on our *pâté de tête*. But no one bought the heart we'd set in the case—which was maybe silly of us, given our particular consumer base. Even the Chapolards didn't do that. Several customers complained about the price of all the meat. Jo and I ended up going home with the hocks and trotters, as well as the skin, bones, and offal, after it all sat in the case for days.

At the end of the week, I asked our bookkeeper whether we'd made any money.

"Are you kidding? We don't even make money when we're just buying box cuts. We probably won't buy whole animals again."

I'd read about a growing number of small butcher shops around the country that claimed to run successful whole-animal programs that sourced from local, humane farms, but I wondered how much of a profit they were turning in the end, and how much product they threw away each week. I knew we hadn't been the most creative with all the parts— and that developing a lasting, sustainable whole-animal program would take time—but even if we had been creative, would heart skewers or smoked ham hocks have sold? I could see how the bottom line at this particular shop would continue to prevent us from experimenting, would force us to supplement with commodity meat just to sell our hundred pounds of sausage a week, which was where the real money was. I wasn't sure I wanted to work for a shop like that. I also had no interest in competitively comparing my soppressata with Jonah's *saucisson*.

At the end of the year, Pastaworks got rid of Jo for reasons that were never made entirely clear. In solidarity, I quit the meat counter, and Evoe soon after.

I didn't regret leaving. But I also mourned the ideal image I'd maintained in my head of a beautiful, friendly, successful butcher shop like the ones I had seen in France, selling every part of the animal, animals that had been raised humanely and with complex flavors and textures in mind, to people who were willing to pay for better meat, who understood the importance of eating the whole animal. We didn't have this customer base, not yet. I'd known this, but, like the farmers I met across the counter at Evoe, had chosen to believe otherwise. I decided to dedicate all my free time to creating this customer base through my classes.

THIRTY-SEVEN

Once Jo and I weren't working together five days a week, once Jo didn't have to deal with my constant pulling her in and pushing her away, I think she realized that her life was much less painful without me in it. We'd loved each other in the way two stranded people clinging to a small piece of floating debris in a lonesome ocean might. We so desperately needed each other to survive, but in the end, we knew that one of us would abandon the other. We'd known this from the beginning, and yet we'd tried everything we could to get around it, to stave off the inevitable hurt. When we finally ended things—our friendship, our affair, our professional partnership—all the disappointment and loss I'd felt in the past two years swung back toward me, but it paled in comparison with losing Jo. It took me so very long to find the right needle and thread with which to stitch up that gaping absence inside of me, an absence that, in the end, I had brought upon myself.

Almost simultaneously, Andrew announced that his work was moving him to Denmark for a year. Before he left, he asked me to marry him. Until then, I'd mostly managed to keep my ongoing difficulties with Jo from him, an intricate mental and emotional feat of compartmentalization of which I was not at all proud. I didn't exactly feel deserving of Andrew's proposal. So, before saying yes, I finally admitted to him as much of what had happened with Jo as I could find words for.

I knew Andrew to be forgiving, generally. Still, I was surprised that he was willing to move forward from this, with me still in his life—whether because he could easily wrap his head around this complexity or because it was all entirely beyond his comprehension, I have never been sure. For a long while after, I'd lie in bed with Andrew and feel the sharp pain of knowing that his presence in my life would always be tied up with Jo's absence. Inside the space between my regret and my relief, I tied a tiny golden weight, such that if I moved in just the right way, I would be forced to remember the profound heft of my own private *and*.

Cleave (1): to divide by or as if by a cutting blow.

Cleave (2): to adhere firmly and closely or loyally and unwaveringly.

When Andrew packed his bags and moved to Denmark, everything went very quiet.

I HELD A FEW CLASSES without Jo, but without her, I doubted myself. I doubted the whole project. She and I had played our respective roles perfectly in sync. At classes, she was the entertainer and I was the organizer. I didn't think I knew how to play her role, too. She was also a cheerleader, not just for the students but for me, and without Jo there to root for me, I began to question whether what I was doing was even meaningful.

In grief, I took a few months off from holding classes, sublet my apartment, and flew to Denmark, where I pitched some stories and continued working remotely for my brother in order to keep paying my bills. Andrew worked all day while I wandered the gray, cold streets of Aarhus, feeling sheepish and morose. I worried that this was the most inopportune time for me to abandon the Portland Meat Collective, but I felt relieved to be far away and anonymous. I didn't want to be seen by anyone. I didn't want my true, disingenuous self to be exposed—the self capable of stringing Jo along for more than a year, the self capable of keeping such an enormous secret from Andrew, the self capable of pretending she was incapable of causing others pain.

But then Kate Hill dug me out of my hole once again. She'd decided to organize a last-minute gathering of women at Camont—Grrls Meat Camp, she called it—and since I was already in Europe, she said, I had no excuse not to come. For the event, she'd invited a small group of apron-clad, meat-wielding, dirt-under-their-nails women from North America to "butcher and bake, barbecue and bonfire" with their French lady counterparts.

And so, in the middle of summer in Gascony, eight of us arrived at Camont with notebooks, butchery diagrams, and knives, ready to collaborate with a group of talented, matter-of-fact Frenchwomen, including the ladies of the Chapolards' cutting room, Jehanne (of course), and several other female farmers and butchers I had not met my first time in Gascony.

Kari Underly a second-generation butcher, arrived from Chicago with her new book, *The Art of Beef Cutting*, tucked under one arm. Cathy Barrow, the brains behind an online DIY-charcuterie-making contest called Charcutepalooza, flew in from Washington, DC, with a suitcase full of her preserved jams. Sarah King, a young, aspiring farmer I'd met in Oregon more than a year earlier, when she hired Jo and me to teach her and her husband, Bubba—they called themselves The Bubbas—how to butcher the pigs they raised, showed up with bottles of Oregon pinot. Barbara Gibbs Ostmann, a seasoned food writer who'd interviewed the likes of Julia Child and Jacques Pépin, made her way from St. Louis. Melora Koepke, a writer from Quebec, drove in from Bordeaux, and Beth Gilliam and Rachael Gordon, two meat-obsessed culinary students from Seattle, made their way to Gascony with suitcases ready to be filled with rillettes and *jambon*.

It seems quite possible that this was the first gathering of meat-obsessed ladies in modern history—I'll just assume the early Paleolithic period probably saw a few chance female meat swaps. Since then, Kate has held a Grrls Meat Camp in other locations around the world each year. And other women-only meat camps have popped up around the country as well.

FOR THE WEEK, some of us stayed in old, funky trailers that Kate had added to her property since I left, while others slept on her barge, in the blue room, and in the Rapunzel-let-down-your-hair room. We took showers outdoors. Kate's noisy rooster, Hank, woke us up before the sun.

For the gathering, Kate wanted each of us to share our various skills. Kari led us in a beef butchery lesson, expertly showing us how to separate the *bavette*, flank, and skirt from the belly, showing us cuts I had never heard of before, the *matambre*, for instance, a "steak tail," and "rose meat." And I nervously led a lesson in pig butchery on a side of Chapolard pork, looking to Kari, clearly the most seasoned butcher in the room, for correction and friendly advice, which she gave me, generously and without judgment. I still had room for improvement, but, remembering what it had been like to stand in Kate's kitchen and break down my first side of pig by myself, I realized I'd come a long way.

The gathering was meant to be for women only, but Kate had a young lad, Dylan, from Ireland, staying with her that summer, and he didn't mind a bit being surrounded by all of us. When we visited Jehanne's duck farm on a day when she had no actual ducks to butcher, Dylan was even game to lie on one of the butchery tables and pretend to be a duck carcass so that, while all of us cackled away, Jehanne could, with her pointing finger, walk us through her basic butchery process.

Daily, we marched into male-run butcher shops and asked them to show us their walk-ins. We drank too much Armagnac around the campfire, seared foie gras for dinner, drank Floc and ate rillettes for every happy hour. Kari regaled us with stories about her first jobs at meat counters. Once, a male colleague threw a knife at her and barely missed. I told her about my struggles with Mike and Jonah at Pastaworks—struggles that paled in comparison with hers—and she nodded her head in solidarity. But she also said she thought things were changing.

"Years ago," Kari told me, "I was the only woman in my business. If I did trainings I'd see one woman in the corner. But somehow I've turned into the Mama Cass of the meat world. Now there are all these young people at the trainings, especially young women."

I'd thought that Jo and I were maybe the only young women in the world looking to learn this stuff, but now, I could see, we were just one of many called to this particular zeitgeist, a fleischgeist if you will.

In the car, on our way to visit a one-woman pig farm, I confessed to Kate all the trouble I'd been through with Jo, whom Kate had met when she and Dominique came to Portland.

"You don't need her," she said. "It's time for you to own this thing yourself. Get on with it. Make it yours."

Kate was right. It was time.

I met back up with Andrew in Madrid, where he'd traveled for work. We ate as much *jamón* as we had room for. I made sure to stand in front of Hieronymus Bosch's *Garden of Earthly Delights* for as long as the Prado would let me. I told Andrew he should come back to Oregon. That I'd be waiting for him. We planned to move in together when he did. And then I returned home to get to work.

Back home, I added more classes to the roster. I held my first pig slaughter class with Levi, using two pigs that Sarah and Bubba King had raised in nearby Newberg. Levi and I held our first rabbit slaughter class. Bob Dickson and I finally held a beef butchery class in one of the cutting rooms of his USDA-inspected slaughterhouse and processing plant, with the explicit approval of his USDA inspector. Our students wore hard hats and hairnets, white coats and rubber boots. Cory Carman, the rancher who supplied us with our carcass, came to tell us about her grass-fed beef operation in the Wallowa Mountains. Bob's employees walked around with meat thermometers and clipboards, monitoring our every move.

I also began to ponder teaching a class on my own, but I worried that, because I didn't work in a butcher shop or as a restaurant butcher, I would somehow be seen as an impostor. Even though I'd written for magazines for more than a decade, I was still hesitant to call myself a writer, so why would I call myself a butcher after only a year of studying the craft? And yet, in classes, I often found myself interrupting the instructors in order to explain more clearly what they were doing, to add more context to the choices they made with their knives, and to urge the students to think about butchery on a deeper level than just *Cut here*, *slice there*. Even if I wasn't the world's most experienced butcher—although I was by that point a perfectly adequate one—I was an articulate one. Perhaps precisely because of my magazine background, I had always been a quick study, plus my verbal acuity and my writer's understanding of the power of story and metaphor seemed essential to the art of teaching. At Grrls Meat Camp, Kate and Kari had urged me to try my hand at teaching, so I took their advice to heart and headed up my very first solo pig butchery class. I loved it. What had I been so afraid of?

On the heels of my first teaching experience, I developed a French-inspired duck-butchery-and-charcuterie class to teach students how to turn one duck into six different recipes, just as Kate and Jehanne did: duck confit, duck rillettes, duck liver mousse, crispy duck breast, duck prosciutto, and duck soup. I thought it would be too obscure for people, but the class sold out in a day. That said, in order to teach my students those recipes, I had to purchase a gallon of duck fat from a source other than my duck farmer. I hadn't yet been able to find a farmer like Jehanne, and so I was still doing what Kate had said I shouldn't do, applying Gascon recipes to non-Gascon ingredients. Still, the students went home excited to try their hand at duck prosciutto and confit. It was a start.

I began to see repeat students in my classes, and a significant number of them confessed to me that they were thinking of quitting their jobs so they could become butchers or farmers. Others told me they were ready to buy a side of pig or a whole lamb and break it down on their kitchen counter. Like proud parents, these students e-mailed me photos of the hams and bacon they hung in their basements. Chefs were starting to sign up, too, wanting to figure out how to incorporate whole-animal butchery into their menus. One meat distributor told me he'd seen a significant rise in the number of orders for whole animals from small local farms since the Portland Meat Collective showed up on the scene. A few guys who actually worked at local grocery store meat counters were also attending classes, confessing that, although they worked with meat every day, they'd never actually seen an entire carcass before.

The Portland Meat Collective was starting to feel more like a movement and less like an experiment. The more the PMC felt like a movement, the more the media came calling. And the more the media came calling, the more I realized that I was becoming a public face for that movement. The more I became a public face, though, the more I began to wonder whether I wasn't becoming a certain kind of spectacle, the spectacle of a female butcher.

THIRTY-EIGHT

A ll right. Are you ready? Why don't you try on this dress and we'll see how it looks on you," the stylist said.

It was a black number, the kind worn to cocktail parties, with a modest cut just above the knee and a gauzy, translucent back.

"And why don't you try on these boots while you're at it." The boots were black, too. Leather. Knee-high. Three-inch heels.

She pointed me to a dressing room that had been fashioned out of a sheet hung from the ceiling. I stepped out of my jeans and T-shirt and into my costume.

The dress was meant for someone who was more well-endowed than I am, but my stylist worked her magic with a few safety pins at the back. The boots were too small for my big feet, but I only had to stand in them for a half hour or so while I posed for the camera.

"How do you wear your makeup normally?" the makeup artist asked me.

"Natural."

"Okay. We'll make it look natural, then."

I looked like a drag queen.

Another dusting of powder on the nose, some mousse worked into my curly hair, another safety pin, and then the photographer.

"Wow. You have *got* to be the *sexiest* butcher I have ever met," he said, and laughed, and then I laughed, and then everyone in the room—his assistants, the stylists, the PR people for the knife company that had chosen me to be a spokesperson for its new ad campaign—laughed, too, in the way that well-meaning people laugh when a well-meaning joke is told to simultaneously call attention to and distract from the fact that something out of the ordinary nags at us from the periphery. It was a gentle, communal, one-of-these-things-is-not-like-the-other laugh.

"So I was thinking you could hold this cleaver up near your face like this," the photographer said, demonstrating. He wanted me to hold it so that the sharp edge was pointed straight at my nose. I gripped the cleaver—it was heavy. The pose felt dangerous, which maybe was the point, I couldn't be sure.

"Like this?" I asked tentatively.

"Okay. Maybe not. How about if you rest the back of it on your shoulder?"

"You know, I don't even really use cleavers," I told him. It was true: I still did not possess the necessary confidence to cleave a straight line like the Chapolards did. If I needed to get through bone, I used a hand-saw, or if all I had was a cleaver, I used a dead-blow mallet, like Adam Sappington had taught me, hitting the back of the cleaver with it as I would a hammer to a nail.

I'd anticipated this moment, and fretted over it, earlier that morning while on the 6 train headed down to Astor Place, where the shoot was to occur. A number of national magazine stories about the Portland Meat Collective had recently come out, and shortly after, a knife company had asked me to take part in an ad campaign. After brooding over it— Would I be selling out? Did I even deserve this?—I signed the contract, and then they flew me to New York City to photograph me before coming out to Oregon to film me. They put me up in a fancy hotel just a block from where I'd worked at *Saveur*. I'd been chosen alongside two men, both seasoned chefs, to take part in a new, ongoing ad campaign, which would continue to feature culinary professionals who were, in the campaign's words, "defining the edge." They agreed to pay me a small amount of money for the work, but it felt like a lot after almost two years of living so close to the bone. I had big, exciting plans to use the sum to pay for liability insurance for the Portland Meat Collective, and maybe a tube of lipstick.

But as the subway hurtled through the dark tunnels underneath Manhattan, I'd worried that if I posed with a cleaver, the image would convey that I was indeed a butcher. A bona fide, skilled butcher. The genuine article. How could I subtly convey to them that I wasn't really a butcher? I was an educator. A thinker. A writer. An organizer. The founder and owner of a unique meat education program. But not a butcher. Definitely not that.

Of course, I had, by this point, started teaching classes myself. In reality, I *did* know how to butcher, I knew how to talk at length about butchering, and I was a damn good teacher. Still, I wasn't sure I wanted to *declare* myself a butcher in such a public way. It was one thing for a magazine to decide, of its own accord, to call me a butcher, but quite another thing for me to sign a contract that had me posing *as* a butcher.

"Okay," the photographer said, "that's great. Now just turn the cleaver a little bit toward me. Gooooood. Great. Beautiful. And now look at the camera, but keep your head facing to the side. Look intense. Look like a butcher!"

A FEW WEEKS LATER, the knife company sent a film crew to Oregon. I drove them to Bubba and Sarah's farm, and they filmed Bubba and me as we roamed the misty pastures, scratching the chins of pigs, feeding them vegetable and bread scraps from a nearby restaurant. They had me simulate a class, in which I demonstrated to volunteer students how to butcher a pig. They filmed me using a cleaver to break down a chicken, even though I preferred to butcher chickens with just a boning knife. Stylists and makeup artists followed me around, and the crew set up big lights in the old house that Andrew and I now shared, a house we were perpetually remodeling together—hanging drywall ourselves, painting our own trim.

When the ad campaign appeared in national food magazines, when my mom called because she'd seen a huge poster of me in a kitchen supply store in Eugene, when an old friend from my magazine days texted me a photo of my painted face on a billboard in Times Square—"Looks like you finally made it in New York," she joked—I felt sheepish. In the ad, the two men had been dubbed "the Rebel" and "the Believer," respectively. I'd been dubbed "the Poet." I found some comfort in this. I didn't write poetry, but I felt I was more of a poet than a butcher, and they'd rightly picked up on the fact that my greatest strengths lay in my ability to write and speak about butchery and meat reform. Although they called me a butcher—"a master butcher"—they also said this: "As a woman in a traditionally male dominated profession Camas is challenging expectations, breaking stereotypes and bringing intellectual depth to the art of butchery."

But all I could think about was that I was a female poet-poser-butcher in a little black dress who wasn't a butcher, and wasn't even really a poet. It was lonely, this thought. I wished I could call Jo—I knew she would help me find the humor in it all—but she had since fallen in love with someone else and made it quite clear that she didn't want to talk to me.

Whether I was right or not, I couldn't shake the feeling that it wasn't my skill as a butcher, but the spectacle of my gender, of my singular and untraditional choice to study butchery *as a woman,* that was garnering me so much attention—not only from this knife company but from the media outlets that kept calling. Were they calling me a butcher because I was one? Or were they calling me a butcher because when they put the word *lady* in front of the word *butcher,* it had a certain sex appeal?

I felt as if everyone had assumed that, because I'd gone and done something so out of the ordinary as a woman, I was also extraordinary enough to master butchery in just two years. I was simultaneously a spectacle and an impostor. I felt *looked at* but not *seen.* It wasn't that I didn't want recognition—when any of us sets out to master a skill, having others recognize our own progress can be gratifying. It was that the recognition—whether in the form of a knife ad or a magazine story— seemed premature, misguided, even empty. All I wanted was to be seen for who I really was: a woman who chose to learn butchery and wanted to continue to learn, a woman who wanted others to learn alongside her. I wanted to be seen as someone in the process of becoming. But where's the sexy headline in that?

At the same time, I knew exactly how the spectacle of my story benefited me *and* my cause. The headlines, my image in Times Square, didn't just fuel the success of the Portland Meat Collective—these accolades were what allowed me to keep learning from teachers and mentors

I wouldn't otherwise have met, which was what I'd set out to do in the first place. I could have refused the interviews, the knife ads, but I didn't.

And yet I kept hearing Mike from the meat counter. "You're not a butcher. You're a meat counter worker." I wasn't even that anymore.

EVEN AS I IMPROVED as a butcher, as I grew used to cutting meat in front of large groups of people and in front of the camera, I found it difficult to judge my own abilities.

You can do this, I kept telling myself. *Just fake it until you make it*. Was I really faking it, though? At some point, didn't I actually know what I was doing?

That it was maybe even acceptable to just be *okay* at butchery, to have not yet mastered the skill *and* receive attention, didn't sink into my brain, either. The only thing that mattered to me was that I was not really a butcher, at least not the kind everyone wanted me to be, the kind who wielded a cleaver with confidence.

It was meant to be flattering, this attention, this modern, post-feminist, you-go-girl praise—I understood that—but ultimately it felt like a rigged game. What I really wanted most was for someone to appreciate what I'd actually mastered and to help me figure out what I might still have left to improve upon. Just because I taught butchery, just because I spoke about it, didn't mean I was done learning.

SHORTLY AFTER THE KNIFE AD came out, the organizer of a Pacific Northwest hunting-and-fishing convention called me up, asking if I'd be willing to do a series of venison butchery demonstrations. I'd never butchered venison, even though my father and grandfather had been deer hunters, but I knew that it would be similar to butchering lamb, which I'd done quite a bit of in the past year, so I said yes. For the demonstration, I stood behind a table surrounded by a sea of people who looked like the people I come from. Hunting caps. Wrangler jeans. Camouflage vests. Work boots. My dad even came to watch.

I was nervous, but I did okay. No one stood up and left in the middle of my presentation, and people came up to me afterward to ask more questions. As I was packing my knives, an older gentleman—blocky, with an intimidating scowl and the sort of thick, hamlike arms and legs that the Chapolard brothers sported—approached me.

"I used to run my own meat-processing facility," he told me in a gentle voice that contradicted his stern expression. "Worked it for fifty years. You did pretty darn good. A lot of great information. Next time, remember to keep your elbows in near your torso when you're cutting. You'll be less likely to get tendinitis that way."

"Thanks for the advice," I said. And I meant it.

"Pretty darn good," he'd said. And he meant it.

"Next time," he'd said, with encouragement in his voice, because he believed there would be a next time. Because he recognized in me room for improvement. He did not seem fazed by who I was. He had not concerned himself with my headlines, my story, my anomalous narrative, my gender. It was a relief, however fleeting, to finally be seen.

THIRTY-NINE

A few short months after the ad campaign's launch, Levi's backyard—the same backyard that had birthed the Portland Meat Collective—became a crime scene.

Sometime between 1:30 and 8:00 a.m. on January 8, 2012, five adult rabbits, including two nursing mothers, and thirteen juveniles vanished from their cages in Levi's backyard. Ten one-day-old babies were left behind and subsequently died.

That afternoon, Levi and I were set to teach a class on how to raise and slaughter rabbits for meat. The rabbits for that class had already been transported to the teaching location the evening before, so they had escaped the hands of kidnappers. Here is how those rabbits died: Within the span of one second, we broke their necks. Within the span of another second, their eyes closed, their nervous system shut down, their brains went dark. That was it. They were alive one second. Gone the next.

This is how the ten one-day-old baby rabbits, without the warmth or sustenance of their nursing mothers, died: Some froze to death. The rest starved to death. A few held on for a while with the help of a man-made nest of blankets and attempts to bottle-feed, but by the end of the day all of them were gone.

WHEN LEVI'S ROOMMATE, Chris, first discovered the day-old babies without their nursing mothers struggling to stay alive, he called Levi, who was helping me set up for the two slaughter classes—chicken and rabbit—that we were scheduled to teach that day. After he hung up, Chris tended to the most important detail: keeping those day-old babies alive and trying to ease their suffering. He was lucky enough to find a local organization, Rabbit Advocates, that had just received a donation of a couple of nursing mothers they were willing to loan him. But in the course of talking to them, Chris realized that those nursing mothers were the very same rabbits that had been stolen from Levi. They had been anonymously donated to the organization that morning and then sent out to various homes to be fostered. When the Rabbit Advocates found out that the rabbits were originally headed for the dinner table, they refused to give them back, including those much-needed nursing mothers.

Meanwhile, the students were already arriving for the chicken slaughter class scheduled to start that morning. Thankfully, Sarah and Bubba had come to help us, so while the three of them began class, I did a little detective work.

A few days before, I'd received a suspicious phone call from a young man who gave his name as Randall Green, saying he was frustrated that the rabbit slaughter class was sold out. "Where will the class take place?" he kept asking me. Something about the guy hadn't seemed right, so I briskly told him I was sorry we couldn't help him and hoped he could take a class at another time.

On a hunch, I went onto Portland Meat Collective's Facebook page and saw that someone going by the name Ron V Green had left a comment in response to our announcement of the rabbit slaughter class: "Shame on the Portland Meat Collective."

Two hours into our chicken slaughter class, Robert Reynolds called me.

Robert had been in the middle of teaching a class himself when a young man walked in, claiming that the Portland Meat Collective had stolen his pet rabbit.

"He's still standing there in my classroom," Robert told me. "He seems crazy and I'd like him to leave."

Levi and Bubba, two large and imposing country boys, jumped into Levi's truck and drove over to Robert's studio. Once there, they encountered the nervous young man, who, they told me later, gripped an empty pet carrier with a paint can inside. He gave his name as Noah Schwartz and said he'd heard a rumor that the Portland Meat Collective "stole people's pet rabbits and slaughtered them for profit."

"If I saw someone hurting a rabbit," Noah said, "I'd hurt *them*." Bubba and Levi stared at him—they really didn't have to do much to look menacing—and then Noah got nervous, jumped into a shiny new Prius, and drove off. Levi wrote down his license plate number.

While Levi and Bubba were confronting the Prius-driving rabbit crusader, I posted on social media that the rabbits had been stolen. Ten minutes later, the first comment I received was from a Ronald Grandon: "I am thrilled that [these] rabbits get a second chance at life, and if true, 10 baby rabbits aren't going to be raised for your sadistic blood lust!"

Through a little more detective work—someone had decided to mess with the wrong ex-fact-checker—I deduced that Ron V Green, Ronald Grandon, Randall Green, and Noah Schwartz were very likely the same person, although I couldn't yet figure out what his real name was. More research revealed that some vegans often used *V* as part of their online pseudonyms. When Bubba and Levi returned, our rabbit slaughter students had already begun to arrive. We had another class to teach. We had rabbits to kill—those we had transported to our class location the night before the rabbit-nappers had shown up at Levi's house. We were all a bit of a mess.

In the middle of class, I received another call from our multinamed friend.

"Is class still taking place?" he said.

"It is."

"Why is it still taking place?" he asked, vocally frustrated.

"Because it's still sold out."

He hung up on me.

Back online, another intriguing post after the phone call:

"Ron V Green," it said, "is disturbed by this 'collective' and would like to see these bunnies thump these fuckers to death!"

The cops had come out to Levi's house to talk to Chris, and I was just about to call the cops myself with information about our new friend when I got a phone call from a reporter at *The Oregonian*, our state newspaper, asking if we'd be willing to comment.

All the media attention I'd received thus far, along with the knife ad, had felt surreal already, but this—which Levi and I later came to call Bunnygate—raised my life in meat to the next level.

Oh, the shit storm *The Oregonian* story unleashed. Within an hour, readers had posted hundreds of comments. Rabbit hunters and those who kept rabbits as pets butted heads, as did meat eaters and vegans, PETA activists and right-wingers. Some people accused us of neglecting to save the baby rabbits—the logic being that we didn't care about the babies, since we were going to kill them eventually anyway—while people who raised rabbits for a living stated that their chances of surviving without their nursing mothers were zero. One conspiracy theorist suggested that we'd made the baby rabbits up to rally public support. We, too, had had trouble believing an animal rights activist—which is who we assumed had stolen the rabbits—would leave ten baby rabbits behind, but the nappers had indeed left them. Had they not seen them? Had they left them on purpose to make a statement? And if so, what statement were they trying to make, exactly?

Just as we were finishing cleaning up, Chris called us, his voice shaking, to tell us that all the babies had died. When Levi hung up the phone, we sat in silence. The babies had been destined for the dinner table *and* the nature of their deaths troubled us deeply. They had suffered greatly, in a way that they wouldn't have if their nursing mothers had not been stolen. They would have had a good life, a good death. Whoever stole them probably believed that killing these rabbits for food was the real crime, that there was no such thing as a good life or a good death for these rabbits.

After the first headline appeared in *The Oregonian*, other stories followed—in *The Huffington Post*, even as far away as *The Washington Post*, as well as on local television news stations. It was all very funny and very Portland to these news outlets—like a sketch straight out of *Portlandia*—but it didn't feel funny to us, especially when I woke up to this e-mail:

> **Things you should do today:**
>
> 1. Get cancer.
> 2. Rot. Slowly and painfully.
> Cunts.

Comments along these lines began pouring in to my Web site:

You are horrible people for doing this. You have forgotten your souls. You have buried your empathy. You have lost what it means to be human and have compassion in this world.

This one kind of stood out:

Killing animals is a sign of developing sociopathy. I would love to kick your ass and make you feel the fear of the animals you think it's so neat to slaughter. . . . I'd love to see you in a lion's cage and see what you think of butchery then. You're Going Down.

A lot of comments relied on a particular brand of circular logic:

People enjoy eating meat. Animals are made of meat. Therefore we must eat the animals. Small woodland creatures such as rabbits, rats, mice, etc. are just the chicken nuggets of the animal world anyway. And yes, I'll eat damn near any animal as long as you cook his ass up right.

There were the patriotic folks, too:

I'll die to protect your right to preach your creed. I'll also shoot the neighbor's dog that raids the chicken coop.

Every once in a while, someone openly grappled with their contradictory feelings:

> **While my husband and I are vegetarians and concerned about animal welfare, we were also saddened that someone(s) hurt your group as they did . . .**

But this was my favorite:

> **I wanted to let you know that even though I'm a vegan, I wholeheartedly support what you stand for. It's obvious that your organization cares about the animals, and fosters . . . respect and humane treatment. So, even though Meat is Murder, I guess you guys are the Dexters of murderers, and I respect you for it.**

MEANWHILE, still refusing to return the stolen rabbits, the Rabbit Advocates hired an animal rights lawyer, whose name—I could not have made this up—was Geordie Duckler. The lawyer told Levi and Chris that the foster parents of these rabbits had grown attached to their new pets and that they wanted to offer fifteen hundred dollars for all the rabbits. (That's eighty-three dollars per rabbit. Levi and Chris typically charged between fifteen and twenty dollars for a live rabbit, which reimbursed them for the cost of raising it. To adopt a rabbit as a pet from the Humane Society costs thirty-five dollars.) Levi and Chris were not interested in making a profit. They were, however, interested in the principle of the entire situation.

"These rabbits were stolen," they told the Advocates' lawyer. "We would like for you to return them. If you would then like to knock on our door and buy them from us at cost, we would be happy to see you."

Chris and Levi waited to hear back. The Advocates waited for them to change their minds. I got my lawyer friend Matt involved. Matt and Mr. Duckler acted as go-betweens.

After a few more days, the Rabbit Advocates returned the rabbits to Levi and Chris, via their lawyer. The handoff occurred in the parking lot of Mr. Duckler's office. According to Chris, the women—curiously, they were all women—who'd adopted the rabbits all showed up in purple Rabbit Advocates T-shirts, with tears in their eyes. The Portland Police Bureau detective assigned to the case came, too, as did a few news channels. Just before Chris drove away with his reclaimed rabbits, however, he realized that one was missing: one of their breeder rabbits, named Roger.

The cops put out an APB and posted it on their Web site:

Update 4:15pm: The Portland Police Bureau has issued a description of Roger the missing rabbit. "Roger is described as small, gray and fury."

Update 4:40pm: The PPB makes a correction: "Roger is described as small, gray, and furry, not fury . . . he is anything but angry."

A day or so later, Levi received a letter from a different lawyer, this one hired by Roger's foster mom. The letter stated that the woman who had "found" Roger had grown attached to him and would like to offer two hundred dollars or the cost of replacing Roger, whichever was greater. In describing her client's feelings for Roger, the lawyer quoted the French writer Anatole France: "Until one has loved an animal, a part of one's soul remains unawakened."

Levi and Chris responded in the same manner: "Return the rabbits, please. Then, just come knock on our door and we'll sell him to you at cost."

Our lawyer drafted a letter in response.

"Your client," the letter began, "has willfully refused to return what is indisputably stolen property—not lost property . . ."

Eventually, the woman decided to return Roger the rabbit to Levi. The handoff happened in her lawyer's office. Roger's foster mother did not show up.

However, a few days later, Levi received yet another letter. Roger's foster mom had found out that Levi volunteered for a charity that took doctors and nurses to Haiti to provide community development services and free medical treatment, and so instead of simply knocking on Levi's door and offering twenty dollars for Roger, the woman offered to donate a thousand dollars to Levi's charity. He accepted the offer and returned Roger to his foster mother. The handoff, once again, occurred in her lawyer's office. It was just Levi, the lawyer, and Roger sitting alone together in a conference room.

"Well, this is awkward," Levi said to the lawyer. He offered her a jar of his homemade rabbit rillettes to thank her for her trouble.

The lawyer handed Levi a check made out to his Haitian charity for one thousand dollars. Rabbit Advocates shortly followed suit, offering the charity another sizable donation in exchange for the rabbits.

In the end, everyone was happy, no one more so than the Haitians, who benefited from a new program launched by Levi, which was seeded by the two generous donations. The purpose of the program? Teaching a community of Haitians—in need of affordable protein in their diets—how to raise rabbits for food.

BUT THAT WASN'T ACTUALLY the end. Shortly after settling with Roger's new mom and the Advocates, a small group of angry protesters with masks covering their faces decided to pay a visit to Levi's house. The same Prius that "Noah Schwartz" had parked in front of Robert Reynolds's kitchen studio was parked on Levi's street.

"We know where you sleep," the protesters yelled into bullhorns. "We will not let you keep your rabbits. We will be back. This is not over. Murderers! Blood is on your hands. You are not men!"

With the help of local authorities and Matt, our lawyer, we determined that Ron V Green, Noah Schwartz, Ronald Grandon, and Randall Green were indeed all the same people and that Ronald Grandon was his actual name. Levi reported seeing Grandon lurking outside of his house several times after that and, once, being followed by him in his car. Levi took to sleeping with a gun under his pillow. Chris, who never locked the doors, began locking them. Nearly a month later, I was still fielding angry e-mails and death threats.

Eventually, the cops advised Levi to apply for a stalking protective order, and we all ended up facing Grandon down in court, Matt acting as Levi's lawyer. Grandon, who had recently been accepted to law school, decided to represent himself, to somewhat humorous effect, although I admired him for his attempt.

Levi, Bubba, Chris, our lawyer, and I sat on one side of the courtroom. Across the aisle, a ragtag group of Grandon's supporters scowled at us. There were maybe ten of them, most in their twenties, with shaggy hair, dressed in well-worn hoodies and frayed band T-shirts. They reminded me of my vegetarian self, in high school and college, young activists in appropriate West Coast uniforms. I felt a motherly protectiveness toward them, even if I also loathed the inflexible dogma they hewed so tightly to.

We could be talking right now, I thought, *finding common ground*. But instead, we all just stayed on our respective sides of the room.

In the end, the judge granted Levi the stalking protective order against Grandon and ordered him to pay our legal fees. Grandon, flanked by two young women he said were acting as his "security," showed up at Matt's office holding an oversize bag full of change and crumpled bills. Matt asked Grandon to count out the nickels and dimes on the conference room table between them.

AFTER EVERYTHING, it was my hunch that few, if any, minds were changed. People who didn't eat meat continued to not eat it. People with bacon fetishes continued to wax poetic over pork belly. Those who think of Levi and Chris and me as sociopaths continue to think of us that way. My classes continued to sell out. Levi and Chris continue to raise rabbits for food. Factory farms continued to cram thousands of animals into confined spaces.

And yet, we all had something in common—Levi, Chris, and I, our students, Grandon, whoever stole the rabbits, and the Rabbit Advocates who took them into their homes. All of us had held the rabbits in our hands at some point. Felt their pulse. Contemplated their lives, their deaths. Each of us had tried to find meaning and make a stand within our very different acts. All of us thought we were doing the right thing.

For those of us who choose to raise and kill animals for food, it's anything but a simple choice. We look the animal in the eye as we breathe the same air. The same tree shades us. The world slows just long enough for us to see our shared place within it. And just before that moment that so few of us want to admit we're capable of—that moment in which one animal chooses to kill another for food—we're not only forced to realize what it means to be human. We're forced to realize what it means to be animal, too.

AT LEAST ONE MIND did eventually change, however. In the years since Bunnygate, Grandon has kept in touch with our lawyer, first sending news that he'd graduated from law school, then that he had passed the bar. Recently, he let Matt know that he no longer identified as vegan, writing that, while he still ate a plant-based diet, he felt that the term *vegan* was divisive and that it prevented him from meeting people at their level.

"It's nobody's business what somebody . . . does with their body," he wrote. And then, "Most of my 'friends' who were with me in the courtroom are no longer in my life."

FORTY

After the protesters and death threats and courtroom drama, I didn't think it was possible for me or the Portland Meat Collective to become even more of a flash point, but the year of the rabbit was just the beginning. If it wasn't yet entirely obvious to me that a lot of people—not just vegans and vegetarians but meat eaters, too—found my particularly unapologetic this-is-where-meat-comes-from stance threatening, the year of the pig head, 2013, made it very clear.

I kicked off the year by posing with a pig head on a silver platter for *The New York Times Magazine*. In the photo, I'm standing at the head of a dinner table. Eight high school students and two of their teachers, all from Oregon Episcopal School, an independent preparatory academy in Portland, sit on either side of the table. In the center we've placed the various parts of the side of pig we've just broken down—chops, hams, shoulder roasts, belly, trotters, hocks, skin, bones—in such a way that the outlines of the pig are mostly discernible. The two students sitting on either side of me gaze at the pig head on the silver platter, their faces a mixture of trepidation and hunger. The other students and teachers alternately eye the pig parts in front of them, some with reluctant smiles on their faces, others looking more somber. A few look to one another inquisitively. Only one student looks directly at the camera, smiling, almost as if he were part of a less serious photo shoot than this one.

The photographer's inspiration for the photo had been Norman Rockwell's iconic painting *Freedom from Want*. In Rockwell's painting, the people at the table appear gleeful, innocent, eager, and also relieved to be free of the feeling of want. One of the major critiques of the painting—a critique that came primarily from outside the United States—was that it glorified overabundance.

For the *Times* photo, the photographer asked that we emulate, as best we could, the expressions of the people in Rockwell's painting. Most of us made it only halfway to those Rockwellian expressions of innocence. Our faces possess hesitation, the good kind. For some viewers of the photograph, the pig on the table might appear gluttonous and overabundant. But to me, it represented thrift, resourcefulness, a tempering of want, a rethinking of need, a willingness to know.

THESE HIGH SCHOOL STUDENTS and I had spent the week leading up to that photo shoot exploring, in proper, hands-on, Meat Collective fashion, the basic processes by which meat gets to our tables, while a *New York Times* writer shadowed us for her story. Each year, students at the school are granted a weeklong, out-of-the-classroom, hands-on learning experience about a specific topic of their own choosing. These students had chosen to learn about meat with me, instead of hat making or dogsledding (two other offerings that year), and on our first day together, I asked them why.

Most of their answers went along these lines:

"I'm here because I love to eat meat, especially hamburgers."

Or "I'm here because I really like chicken."

Most of the students had seen the workshop as an opportunity to eat a lot of meat and maybe learn how to cook it. My secret hope was that, through this class, they'd come out the other side asking questions about their very desire for it.

I started the class out with burgers.

"Where do burgers come from?" I asked the students.

"Meat," they said.

"What kind of meat?" I asked.

"Beef," they said.

"What kind of beef?"

"Ground beef!"

"Where does ground beef come from?"

"Cows."

"What do we know about those cows? How were they raised? What did they eat? How old were they? How were they killed? How do you get ground beef from a live cow? What part of the animal does it come from?"

Silence.

We brainstormed what we knew: that most of the hamburgers we ate probably came from larger factory farms or confined animal-feeding operations—I led the witnesses only a little, providing them with statistics and other factual evidence to inspire their answers. We brainstormed what we collectively knew about factory farms: "cramped," "gross," and "stinky" were words the students threw out.

I asked them if they had ever heard the term "grass-fed" or "grain-fed," and then we brainstormed what we thought we knew about these terms.

"Don't all cows eat grass?" one student asked.

"At the beginning of their lives they do," I said, "but most of them are eating grain by the time they're adolescents, even if grain isn't something they can easily digest."

I gave a brief overview of diet, breed, exercise, muscle definition, and the various legal and informal definitions of *humane*, and then I drew a butchery diagram of beef.

"What part of the animal do burgers come from?"

Silence.

"Can we even know for sure? How might we find out?"

This is how most of our conversations went. I asked them what they thought they knew and what they were even allowed to know. Then I asked them what purpose it might serve not to know, and urged them to figure out how they could know more.

We made two different kinds of burgers the first day: one using pasture-raised, grass-fed beef, and one using factory-farmed, grain-fed beef. To make our burgers, we ground the meat ourselves, first examining the difference in muscles from different parts of the animal. We inspected the fat, comparing the color and texture of each.

As the ground meat and fat coiled out of the grinder, a few students closed their eyes in disgust. "Gross!" one of them said.

"Let's talk about 'gross,'" I countered. "This is going to be turned into a burger that you will gladly eat. Will it be gross then? If not, why?" I asked them as many open-ended questions as I could, in order to get them to switch on their brains and think.

When we tasted our burgers, I asked them if they could taste the difference. Most preferred the grain-fed, factory-farmed burger.

"Why do you like that one better?"

"Because it's what I'm used to," one student said.

"Now that you know how that burger meat was raised, does that change anything?"

The students blinked at me and said, "Maybe."

ON THE SECOND DAY, we visited Bubba and Sarah's farm. The students met the pig that we would be butchering later in the week and chose to name her Wilburess.

Bubba showed them what she'd been eating for the past seven months, and how much space she'd had to move around in the pasture. He encouraged them to pet her and to feed her pieces of raw squash. We led the students—in their skinny jeans and hoodies, clutching their cell phones, texting their friends selfies with the pig—to the barn, where a sow had recently given birth, and they cooed over the fuzzy baby piglets.

I felt like the bearer of bad news, but as the students reached into the pen to pet the piglets, I reminded them that Bubba would be taking Wilburess to the slaughterhouse over the weekend, and that on Monday we would witness the slaughter.

"Do any of you have any questions or concerns before we go on Monday?"

"Do we have to watch the slaughter?" one girl asked.

"It's your decision. We aren't going to force you. But the idea is that we're going to show you every part of the process of how meat gets onto your plate. We're going to show you the most ideal version of that process, but the process isn't always as ideal. Remember, not all pig farms look like Bubba's, and not all slaughterhouses are going to look like the one we go to on Monday. I'm not here to tell you how to feel or think.

But I guarantee you that as you go through this process, you will never look at meat the same way again."

"Can we take pictures?" a young freshman boy asked.

"No," one of their teachers said. "We're all going to leave our cell phones and cameras at home."

"Let's respect what's unfolding in front of us and try to be present for it," I said.

I wasn't sure whether the students, all coming of age in the era of Snapchat and Instagram, even knew how to do that, but I hoped they would figure it out.

A FRIEND WHO WORKED at the slaughterhouse had swung a visit to the kill floor for us, convincing the suspicious plant manager that my intentions were pure. In the parking lot of the slaughterhouse, the manager laid down the rules. *No cell phones. No cameras. Wear protective clothing at all times. Stay quiet. Follow my directions.*

"None of you are journalists, right?" he said, chuckling afterward at the absurdity of such an idea.

After the students booted up and donned their hard hats and white butcher's smocks, the manager led us, single file, into a viewing area to the side of the kill floor. Between the noise of the hoses they used to spray the carcasses down, the continuous air-pump sound of the bolt gun, the music blaring from the radio, and the hiss of the blowtorches used to scald the hair off the pigs, we couldn't hear the plant manager very well, so the students, suddenly appearing so much smaller in stature than they had in the parking lot, stood there watching and sniffing the air, without much in the way of a translator. I watched the students watching. Their faces were straight, somber, serious, which is not to say

I couldn't also see empathy on their faces. At the same time, none of them turned away at any part of the process. No one joked with the people next to them or looked the other way. They were all clearly present.

As we filed out, the students remained quiet—a stark contrast to their usual joshing around with one another.

Once in the parking lot, I asked them how they were doing.

"I was surprised at how I felt," one said. "I don't have any words for it."

"I felt sad," one girl said. "But it also wasn't as horrible as I thought it would be."

"I imagined something different. It was almost too calm," another said.

"I'm not sure if I will eat meat again or not."

I urged them to try to find words for it if they could that night. And I told them that in two days we would be butchering our pig carcass and making sausage and bacon and ham, which they would all go home with to feed their families.

WHEN THE STORY came out in *The New York Times Magazine,* the backlash was immense. I spent another month fielding angry e-mails and death threats by phone. (On the other hand, my classes also sold out much faster than usual.) "Disgusting" and "sickening" were the words of the hour from online commenters, many of whom said they ate meat. I was getting tired of playing nice.

Parents argued that it was inappropriate to teach young children to be "violent" and to "objectify" animals. I was accused of teaching children "disassociation," when, in fact, association had always been my goal—association that forced all of us to grapple with how to eat and be in the world.

One of my students responded to the angry online comments.

> As a student who actually partook in this weeklong class, I can say with certainty that it has changed my views on meat and the meat industry. Am I a vegetarian now? No. I will still continue to eat meat, but now I eat meat a little less frequently, and always in awareness of the means that got it to my table. . . . Also for those criticizing the school, please understand that every student who participated in this class voluntarily signed up for it. I personally signed up for it because as a meat eater I felt it was important to understand the whole process and have a greater understanding of meat beyond the packaged meat that so many people buy from supermarkets every day. If anyone has questions, please ask! I'm happy to share my experience with you!

Not a single person responded to him, but I was proud of him anyway. No amount of dogmatic thinking or online shaming would ever take away my dedication to those students and their grappling.

FORTY-ONE

At the tail end of the year of the rabbit, I ended up writing an essay about Bunnygate for a local humanities journal, and the story got picked up by the public radio show *This American Life*. As a result of that story, people from all over the country began contacting me to ask if I would start a Meat Collective in their community. In the year of the pig head, I open-sourced my educational model, raising money on Kickstarter to help individuals in Olympia and Seattle start their own collectives, and then launching an entire nonprofit, the Good Meat Project, whose mission is to spread my educational model across the country. Andrew and I were also planning a wedding for that summer, which made the year of the pig head an incredibly busy year, especially once Martha Stewart's people called me.

"Congratulations," her editor in chief said when I answered the phone. "Martha wants to give you an American Made award and fly you to New York for the ceremony."

"A what?" I said. I'd never heard of this award.

She went on to explain that Martha had chosen to give the award to ten entrepreneurs who "showed innovation and creativity in their respective fields." We'd be featured in the magazine and in online videos. *How did Martha even know about me?* I wondered.

A few months later, Martha sent a film-and-photography crew out to Oregon and asked me to stage a class, so that they could better control lighting and props. For the shoot, I had about six volunteer students gather around me at a table as I worked on a side of pork. About halfway through, the photographer's assistant pulled me into another room to ask me if I could get two of the students out of the picture, in a way that wouldn't seem obvious.

"Why?" I asked.

"We're just going for a certain look," she said. "And some of them don't have it."

Right. This was Martha. Everything needed to be just so. But I made the photographer's assistant figure it out. We didn't discriminate in my classes.

After the students left, I posed for the camera with various knives, and the ever popular cleaver, of course. Then they had me lay out different cuts of meat for them to photograph.

"Why don't we start with recognizable cuts," the photographer said. "Like pork chops, and maybe bacon."

I put a piece of belly on the table in front of him.

"Could you maybe trim some of that fat off? And is that skin? Maybe trim that off, too."

"But bacon is made of fat."

"Yeah, but our readers are into healthier living."

"I thought you wanted the cut to be recognizable as bacon?"

"Okay, that's fine, let's just take the skin off, then," he said.

"Why the skin?"

"Because people won't like it. They won't know what it is."

"But that goes against all the reasons I do what I do."

"I know, but it's just not appetizing," he said.

"But it is! I just made a Thai sausage with pigskin in it the other day," I said.

"Right. Well. I'll just shoot the pork chops, then."

Enough with the pork chops, I wanted to say. *Every time you show a picture of pork chops, you convince more people that pigs are made of nothing but pork chops. This is why we have to produce so many pigs, in the worst of ways—to satisfy all the people who want those pork chops. Let's show the pig head. Let's show the tongue.*

As if reading my mind, the photographer turned his camera to the pig head.

"Hmm. Could you maybe wrap that head in butcher paper but still make it recognizable as a pig head?"

"Why not just photograph it unwrapped?" I asked.

"I just don't think Martha's readers will be able to handle it," he said.

I sighed. I wrapped the pig head with a few long pieces of butcher paper and some masking tape. It looked like an unidentifiable, crinkly white meat football.

The photographer's assistant frowned.

"Hmm, let me take a stab," she said, unwrapping the pig head and starting over. She proceeded to work her magic, wrapping the white butcher paper just so around the snout and the cheeks so that the paper clung tightly to the contours of the pig head. It looked a little more like a semi-realistic, albeit artistic, rendering of a pig head instead of a football. "Now we just need some toothpicks to make the ears stand up," she said.

I laughed. But she wasn't kidding.

"Could you just hold the pig ears up for me?" she said as she poked toothpicks into them to prop them up, and then wrapped paper tightly around each of the ears.

"There!" she said. "What do you think?"

In the face of this utterly astounding bit of porcine origami art, I had no words. Martha had awarded me with publicity, a free flight to New York, a free hotel room near Bryant Park, a fancy reception in Grand Central Terminal, a party, a tour of her magazine offices, and a feature in the magazine, because of my proven "innovation and creativity." But here we were masking the very things my innovation and creativity had been in service of. Here we were covering it all up with white butcher paper.

"It looks good," I said, and then grabbed a Sharpie and wrote PIG HEAD on it.

EARLIER THAT YEAR, at the request of many of the farmers I now worked with, I'd taught a pig-head butchery-and-charcuterie class to twelve students in Portland. The stated goal of the class was to teach people how to turn every part of the head into food. My unspoken goal was to force people to confront the fact that meat comes from animals, the head and face being the most visceral reminder of that.

For the class, I taught the students how to remove all the meat, fat, and skin in one piece, how to extract tongue from skull, how to season everything correctly, and then how to roll and tie it all together and, through roasting or poaching, transform that into a rich and hearty *porchetta di testa*—an Italian-style head cheese—that you could then slice and serve with mustard and crusty bread for an appetizer or as an accent to a meal.

A reporter for a local alternative newspaper decided to write about the class. "Barnyard Butchery: The Decadence and Horror of Butchering a Pig's Head," his headline read. The use of the word *horror* made me wince, because it would surely elicit a response, but I liked the adjoining *and*.

"If there are five basic food groups," he wrote, "there must surely be more that are too cerebral and unquantifiable to make it into the federal labeling process. I often like to say pleasure is a food group we shouldn't discount; this [class] carved out new territory beyond that. Or rather, it exposed long-buried territory we see only flashes of when we get our hands dirty and cook: involvement, pride, and understanding. It also, in a decadent fashion, was a challenge, a risk, and a thrill, made all the more satisfying for its sensual reward."

One reader, a meat eater, just didn't get it.

"Let me start off by saying that I eat meat," she wrote in a letter to the editor. "Let me add that I don't think that working in the meat department at a grocery store makes you a bad person by any stretch of the imagination. If my neighbor prides himself on BBQ ribs, more power to him (and pass the cole slaw). I have known many fisherman [*sic*], but none made a fetish of the entrails.

"At the same time, making a fetish of stripping the meat from a pig's head as art, entertainment and as a public act is just plain creepy. Not dissimilar from her public butchering of animals widely considered as pets."

The author went on to suggest that "normal" hunters or backyard chicken raisers wouldn't ever do such a thing, and that while there were things we had to do in this life, like kill mice, we shouldn't normalize "getting into death for fun," as I was apparently doing by teaching people how to turn a pig head into food.

She signed her note "Oregon Mamacita."

How I wished Oregon Mamacita would sit at my table and let me cook for her. We had so much to talk about.

In my world, I'd tell her, a normal hunter who takes the life of an animal should feel a responsibility to turn the entire animal into food. In Oregon Mamacita's world, a hunter, apparently, should kill an animal, use only some of the cuts, and leave the rest for the birds, because eating any other part would apparently be creepy.

In France, the people who worked at meat counters cleaved pig heads in half every day, respectfully, in the name of thrift and sustenance, and turned them into *pâté de tête*. Yet I hadn't heard anyone accuse these butchers of "making a fetish of stripping meat from a pig's head as art and entertainment."

In Gascony, Kate and her neighbors and friends taught me that when your chicken got too old to lay eggs, you turned her into *poule au pot*. You'd maybe throw the iron-rich heart onto the grill or into a confit, and turn the liver into mousse. To Oregon Mamacita, any backyard chicken raiser who used "entrails" for food was not normal.

How did we get here, Oregon Mamacita?

And she was hardly alone. Oregon Mamacita was the kind of reader Martha's photographer had in mind when he asked me to wrap that pig's head in butcher paper.

IN THE END, Martha ran the pictures of pork chops in the magazine. The pig head never made an appearance.

Before the awards ceremony in New York, each awardee was asked to pose on the red carpet with Martha, after she posed with actual celebrities, like Christie Brinkley and Bobby Flay. Cameras flashed. A crowd gathered. I kept turning to my fellow awardees to see if anyone but me felt at all out of place—I mean, where had Christie Brinkley even come from?—but no one gave me a knowing look back.

When it came my turn to pose next to Martha, she kept her face pointed toward the cameras and didn't look at me.

"Look straight ahead," she said. "Keep smiling."

As we posed and smiled and looked straight ahead, she talked to me out of the side of her mouth.

"So you're the butcheress," she said. "Most people don't know this, but I come from a family of Polish butchers. You want to know the best way to kill a turkey?" she asked me. Thanksgiving was right around the corner.

"Tell me," I said.

"You feed it vodka first." She looked over at me with a defiant look in her eyes and winked.

I wondered if those words, let alone a pig head, would ever grace the pages of her magazine. I understood, finally, that including me in her magazine at all—posing with a cleaver and those tender, mild, inoffensive pork chops—was something of a risk for her, a risk she might not have taken four years earlier, when I went to France to learn how to kill my own dinner, but which she was now willing to take. I liked to think that I'd helped to make that possible.

Before I held my first Portland Meat Collective class, Jo had asked me what my long-term goals for the project were. I told her that I hoped someday the Portland Meat Collective wouldn't even have to exist, because everyone would already possess the knowledge that our classes offered.

"That's a long game you're playing," she said.

It was indeed. It still is.

FORTY-TWO

It has been nine years since I first went to France to learn how to turn a pig into pork chops and a pig head into pâté. My classes still sell out. I have competitors now, other people who see the need for such education, which means that the long game hasn't gotten any shorter. Occasionally—usually when Levi and I hold a rabbit slaughter class—someone threatens to protest. I always call Levi up to warn him, but we never cancel.

As I write this, Kate and Dominique are heading to Oregon to teach with me, as they have every other year since I went to France. But this time will be different. Last year, a few months before Dominique retired, with plans to take his pig show on the road with Kate, his doctor diagnosed him with a serious form of cancer. When he comes to Oregon, we will gather as many students as we can who passed through the doors of the *salle de découpe* after me. We will butcher and cook and eat with him. We'll show him what we've done with all that he and his family taught us—the butcher shops and meat schools and full-circle pig farms we've each started. He will remind us to stand up straight, to breathe, and to smile. *Tout seul, tu meurs,* he'll say, while slicing his formidable *jambon* for us to taste. We are a tribe now, all of us who dropped whatever it was we used to do, whatever it was we thought defined us, stepped into that black hole, and decided to linger there long enough to find the meaning we'd each been searching for.

Every year, Kate holds a Grrls Meat Camp somewhere in the world. When I am able to attend, I find myself surrounded by women who, like me, also dropped almost everything sure and reliable in their lives in search of the real thing, the genuine article. We're also a tribe. When I say my name is Camas, they ask me for my story. When I am done telling it, I always ask them for theirs.

A PHOTOGRAPHER CAME to my house recently. Studying me through her camera lens, she probed for more information about the classes I offered, but when I began explaining, she interrupted.

"Actually," she said, "don't tell me."

I laughed, but she wasn't joking.

"No. Really. I mean it. I just don't want to know," she said.

"I don't ever want to know what that feels like," I said to her. We didn't even bother to make small talk after that.

EVERY ONCE IN A WHILE, after I have hung up the phone from my once-a-year catch-up with Jo—all we ever really allow ourselves anymore, since she moved north with her wife to start a new life in a different fertile valley, since she is busy with her new daughter and I with mine—I ask Andrew whether he wishes I'd shared with him all the lurid details of my relationship with this woman I once loved. Sometimes I think I want him to know.

I don't really need to, he usually says.

I like to think I am the kind of person who will tell him anyway. But

I do not. In truth, I am thankful that I will not need to go digging for new words to explain that unseen road I entered upon so long ago.

SOMETIMES, when I am at the dentist or the doctor or making small talk with a stranger at a party and they ask me what I do, I do not tell them the full story. "I'm a writer," I say. Because it's easier that way. Because most people can conjure a story or a picture in their head when I say this. Sometimes I don't want to risk the stories they will tell themselves if I say, "I am a butcher, but not a butcher, not really." Sometimes when people ask me my name, I just go ahead and tell them it's Jennifer.

THERE ARE SO MANY THINGS we do not want to tell. So many things we are not told. Often, this is out of convenience. It's just too long and arduous to tell, after all, and that's all the not telling is really about. Who has the time these days? Besides, sometimes people just don't want to *be* told, so the feeling is often mutual.

Or maybe an adult thinks a child is not ready—and maybe, in fact, the child really isn't. Or a writer decides not to tell the whole story— just the parts that seem most relevant to drive home her point. Or a politician thinks his or her constituents are not prepared to swallow the truth, because the truth often involves sacrifice, which in turn agitates the populace, and so the politician tells some of the story but covers other parts, to minimize the risk. And, sometimes, don't we know this is happening but just let it go? Isn't there something about not telling that we are all complicit in?

There are, of course, more generous ways to look at not telling.

After all, sometimes the not telling keeps us always wondering, which may be a delightful and flirtatious way to move through life, but only if someone throws you a few bread crumbs every once in a while. A literature professor in college once suggested that Victorians, while seemingly tight-lipped and buttoned-up, swathed as they were in all that tucked garb with one hundred clasps, were masters of the slow, controlled reveal. The mere appearance of a bare anklebone sent them all aquiver.

Plus, whole families, governments, corporations, and cultures—not to mention our entire industrial food system—are built on the tenets of not telling, so one might argue that not telling is important to the stability of our lives. But I have often not told, or not wanted to be told, out of a false sense of stability, and maybe also out of laziness or fear, or both. And lately, I worry that not telling—and, thus, not knowing—will, over time, dull me into thinking things are simpler than they really are.

"Sometimes things just really are simple, sis," my twin brother, Zach, often says to me, which is not at all to imply that he is in any way simple, or simple-minded—he is, in fact, quite the opposite. But still, he has asked me more than once, "Why do you always have to process everything?"

By *process* he means, I think, an insistence on revealing everything at every given moment. *Everything out on the table. Empty your pockets. I want the truth, the whole truth, nothing but the truth.* An insistence fueled by a persistent belief that, always, there is something that has not yet been revealed, which maybe implies that there is no such thing as *everything out on the table.* Because, even once you've emptied all your pockets, I'll still squint at its contents, all spit and shine under the bright lights, and say, *I don't believe you. Is that all there is? Show me something else.*

STILL, I find myself not telling when I think that the telling will be too hard, that it will require too much of my energy, and too much of anyone else's. I wish, out of empathy, to prevent others from bearing the weight of it.

A student once told me that when her grandfather found out she'd attended a pig slaughter class with us, he'd been disappointed. "We worked so hard so that your life would not be hard," he said. "Why would you want to go and make things hard again?"

And it's true. They did work hard to make our lives easier. They fought in wars. They lived in mud huts. They traveled in rickety wagons through snow and sleet. They made grave mistakes. They did terrible things sometimes. They also invented dishwashers and refrigerators and vacuuming robots and spaceships and cars and mechanized slaughterhouses and Styrofoam and underpaid migrant labor and bubble wrap and Scotch tape and electric chain saws and pull tabs and vitamins. And frozen chicken nuggets, which are so much easier than having to kill and butcher and fry up a chicken all by your lonesome.

A FRIEND'S THERAPIST recently said to her, "You know, you don't have to be honest all the time about everything."

Imagine that.

My friend went on to explain that her therapist had meant she didn't have to bring up every difficult thing with those around her all the time, because not everyone wanted to talk about *everything*, especially if *everything* was too hard.

My friend and I, both being the sort of people who have a tendency to want to wave our hands above our heads and point and yell, "Look! Look! There's an elephant in the room! An elephant!" on a somewhat daily basis, went back and forth about it for a while and concluded that not telling, not revealing, was simply inefficient. Because whatever hard thing you're working so hard to make sure no one knows, or to make sure you never have to know, will always catch up with you—and them—eventually. And then, when the catching up occurs, everyone's freaked out, screaming, "How could you! The horror! What happened? How did I never know this? Why did you not tell me! I don't want to know!" It all seemed like a lot of wasted energy to us.

I think, in our own stubborn, idealistic way, my friend and I were arguing for a kind of communal willingness to grapple. *Come out, come out, wherever you are! Let the silver light in! Let us see you!* But we were also maybe being absolutist brutes, flattering ourselves into thinking we could handle being told just about anything at any given moment. In truth, neither of us has ever been that strong.

ONCE, WHEN MY NIECE, Georgia, was quite young, she asked her parents a very important question. She asked it at night, just before going to sleep. It had been a long process getting her into bed. There was the brushing of the teeth, and the washing of the face, and then the brushing of her stuffed rabbit Bobby's teeth and the washing of *his* face. And then the taking off of the day's clothes and the putting on of pajamas. Then Bobby's pretend pajamas had to be put on. A few pages from the latest chapter book read out loud. A song or two. Then the dimming of the lights, but then she was out of bed again to check on an empty shelf just to make certain it was empty. Then the tucking in again.

Another song. When Georgia finally felt safe and warm and settled enough to begin sucking her thumb and close her eyes, my brother and sister-in-law kissed her on her cheek and smiled at her. "Good night, Georgia. We love you."

As they pulled the door closed so that just a crack of light from the outside world still shone into her dark room, Georgia called out.

"Papa?"

"Yes, Georgia?" My brother opened the door and poked his head back inside to look at his daughter.

"What's this?"

"What's *what?*" my brother asked.

Georgia sat up in bed and extended each of her arms out in front of her body on either side as far as she could and, scanning with her wide-open eyes the entirety of the space between her hands, she said, "All this. What's all this?"

My brother sighed and sat back down on her bed. "That's going to take a long time to explain," he said, smoothing back her hair with his hand.

This was probably eight or nine years ago, but sometimes, when I have not seen my brother and sister-in-law for some time, and become frustrated with our lack of contact, I try to remind myself that they are probably still sitting at the foot of my niece's bed explaining *all this* to her, and so a phone call or a quick glass of beer are simply out of the question. Meanwhile, they both well know, Georgia will already have burst forth into the real and true *all this*, with arms and legs swinging.

This is probably not my story to tell, but I tell it anyway, often, because I like how it forces those in the room, me included, to try to remember what it was like to be on the other side of *all this*. Not to imply that we master it by the time we become adults, but there comes a point where you step into *all this,* and you can never back your way out of it again, at least not in the same direction from which you came.

It reminds me, too, how very difficult it is to remember correctly, let alone comprehend, the person we were before we stepped into *all this.* And, really, *all this* isn't even just one realm. It's a multitude of realms— hidden chambers of the heart, expansive black holes, vast pastures, pot-holed dirt roads—which means there are a multitude of selves whom we can hardly comprehend we ever were. I'm not so interested in *If I knew then what I know now.* I'm interested in *How I might ever know again what it was like to not know then.* There is really no such thing as a tell-all, in other words.

This may all be self-evident, but when someone sits down to relay the story of who they were before they entered some kind of midlife al-ternate version of my niece's *all this,* complete with sharp knives, cum-bersome cleavers, and those always problematic pig heads, they are not only committing themselves to sitting at the foot of a reader's bed in order to explain *all this,* for as long as it takes to explain and no matter how difficult. They are attempting to expose the complex relationship between known and unknown, seen and unseen, told and untold. They are folding experience back in on itself, "comparing like with unlike," as John Berger puts it, "what is small with what is large, what is near with what is distant." They are, in fact, risking severe temporal hubris.

But here I am anyway, stretching my arms out in front of me and staring at what lies between. I forgot to mention that I am known for my irregularly long arms. When I'm standing up straight, the tips of my middle fingers reach just about all the way down to my knees.

"It must have been so disturbing," so many people say when I tell them my story.

"It sounds so hard."

"Why would a girl like you go and do something like that?"

During those first few weeks in France, Dominique instructed me to open the pig like a book.

He meant this literally, of course. You run the tip of your knife underneath the rib cage to release the fascia and fat that adheres this elongated and warped fence of ribs to the shoulder and belly. Then you set your knife down and, with your own two hands, you open the rib cage away from the pig's once alive, now dead body in the same motion you would use to open one of those big, old hardcover tomes that need their own pedestal to rest on.

I suppose I took it literally, too.

As in.

This pig.

A book.

All this.

A telling.

A NOTE FOR ALL
THE SEARCHING PEOPLE WHO HELPED
TO MAKE THIS BOOK HAPPEN

I swore I wouldn't write about my time in France, but the story eventually got the best of me and I am thankful that it did. There I go thanking the story. Who does that? I do. But also, during my first few weeks in France, Colman Andrews dropped me a note, told me as much would happen, and encouraged me to at least think about writing a book about it. "I'm not ready," I said, but I secretly thought about it for a long time, so thank you for planting the seed, Colman. Thanks, too, for bringing me into the fold at *Saveur* so long ago, because the fold—Margo True, Melissa Hamilton, Kathleen Brennan, Caroline Campion, and Kelly Alexander—taught me that you can make food writing do anything you damn well want it to.

Once I decided I was actually ready to write the book, Emma Parry, my agent, expertly shepherd me through the process. Thank you to Jessica Applestone for introducing us. Emma not only helped me to give shape to what was then a rather dense *brouillard* of an idea at best. She helped me, in the most seamless and graceful of ways, to find the book its proper home here and abroad. Emma also repeatedly assured me, after many panicked calls—*yeah, no, this whole writing-a-book-thing is not for me*—that I was entirely capable. I can't really imagine writing another book without you, Emma.

Ann Godoff and Scott Moyers, as soon as I sat down in your office and Scott pulled out that crinkled piece of paper from his pants pocket with his chicken scratch on it, smiled, and said to me, "This is what I think your book is about," thus causing Ann to adjust her glasses and look up and to the left in a display of deep curiosity—I knew you were the ones for me. Thanks to you both and the rest of your team at Penguin for seeing the project clearly from the very beginning. Thank you, especially, to Scott and his crack editorial team, Christopher Richards and Kiara Barrow, for guiding me when I could no longer see.

Also, across the pond at Picador, my British editor, Sophie Jonathan, pushed and prodded my text in all the right ways and the book is all the better for it.

Books take way more time than I thought they would and finding time is not my strong suit. Luckily there are organizations whose sole purpose is to give time to writers. Thank you to the Logan Nonfiction Program at the Carey Institute for Global Good for accepting me into their nascent residency. Thanks especially to fellow residents Justin Cohen, Matt Young, and Susannah Breslin for reading my first sloppy attempts. An underwater photographer named Jason weekly delivered firewood to the front door of my cabin at Caldera's artist residency. That pretty much sums up why I love Caldera. Also, thanks to fellow resident Kim Calder for her sharp editorial eye and keen intellect. Lastly, the good folks at Willapa Bay AiR plied me with spirited conversations and delicious food while I wrote the hard stuff.

Frances Badalamenti, thanks for letting me turn your coast cabin into my regular writing retreat. It's quite possible that most of my book came to me in the form of sauna sweat. Also, thank you for your early input and insight into my first chapters.

Zach, Aimee, and Georgia, thanks for letting me take over your basement office when things got hairy at the end and thanks for offering the kind of family support a girl needs in times like those.

My Good Meat Project board members kept the nonprofit flame going while I was busy with my head in the book. Thanks for your patience and for your continued enthusiasm for this constantly morphing project. Tanya Harding, Sarah King, and Sarah Wong—my pioneering board members—thank you for your continued belief in all of this, and in me.

Thanks also to the people who made the Portland Meat Collective run without me while I was off in the literal and metaphorical woods writing, especially Faye Holliday and Beth Collins—plus, all of our PMC instructors and class assistants who have kept the PMC running like the well-oiled machine you have helped it to become.

The tricky, shifty nature of facts—everything is complicated once you dig even a centimeter deep—keeps me up at night. Thanks to Sylvie Lubow and the folks at Unfurl for helping me sleep better. Thanks, too, to Dr. Michele Pfannenstiel, Dr. Andrew Milkowski, and Adam Danforth for vetting all the complications that worried me.

Many thanks to Emily Chenoweth, BT Shaw, and Joe Streckert, who read their work at Livestock, an event that occurred in Portland in 2009, and whose works from that event are quoted in the book.

I've been lucky to have stumbled upon so many generous mentors over the course of my life. If I were to draw up a family tree of all of them, the roots of that tree would take me back to Tim Goss (aka Mit Ssog) for teaching me, at a young age, how much more enjoyable it is to live a questioning, curious life, and to Marie Pickett who encouraged me to apply the wilds of my imagination to the wilds of Fruitway Road.

The number of farmers, butchers, and chefs I have learned from over the years is too numerous to list here, so let's just say that if you have ever come into contact with the Portland Meat Collective, these pages exist because of you.

Robert Reynolds, you were supposed to be around to read this. Yet, at times, I sense you standing behind me, reminding me the commas only matter if I tell them to *do* something poetic, and reminding me to drink my daily dose of bubbles.

Tom, roaster of meats, you dwell in the spaces between each line of this book.

Jill Davis makes me jump out of planes and I love her for it. 'Nuff said.

I can't really imagine writing about the world around me without Robin Romm writing somewhere nearby. Remember when we were eighteen and we moved into that cramped, rotting silver bullet trailer together up the Siuslaw a few miles from the Pacific, determined to live the young poet's life? That's when I knew you were the writer for me. Here we are now, still, thinking hard together about the hard stuff.

Dad, thanks for teaching me to gut my first fish. The lessons you taught me early on about the real world of real things clearly stuck, even if I forgot them for a while.

Mom, thanks for inspiring me, probably without even meaning to, to question those fish guts. Also, admit it. Without the great ongoing pork chop debate, our relationship would be so boring, and, I'd argue, so would this book. Thanks also for taking me to the library every weekend when I was a kid, for teaching me the intimidating power of the red pen in my adolescence, and for letting me, always, curl up next to you to read a good book.

T. S., it was all worth it. I warned you I might write about it some-day. You said, *I trust you to put the right things in there*. I hope that I did.

And.

A. R., how is it that you make everything that feels so hard—everything that finds us with our hearts hanging out—all, somehow, mysteriously, seem so easy. I know you know how much I owe you. Because, math.

Finally, I will forever be indebted to Kate Hill for so generously in-viting me into the life of abundance that she has chosen to create and to Dominique and Christiane Chapolard—along with the rest of their family. With you all in the world, one can never really be alone.

A NOTE FOR OTHER SEARCHING PEOPLE

For anyone wishing to learn what I have in the past nine years, there are so many more resources now than when I set out—too many, in fact, to detail here. I do, however, maintain a running list of these resources on the Portland Meat Collective Web site (pdxmeat .com/resources). Go there and ye shall find a good starting point in the form of books, films, organizations, charcuterie workshops, annual meat gatherings, Web sites, podcasts, and more to guide you on your journey.

To find out how to start your own Meat Collective, or to locate other Meat Collectives that have launched, visit the Good Meat Project Web site (goodmeatproject.org). If you are a farmer, butcher, or a chef looking for Meat Collective–style educational opportunities geared toward your needs and interests, you'll find information there as well.

For all my fellow meat ladies out there, check out Grrls Meat Camp at grrlsmeatcamp.com. I'll see you there.

Kate Hill still welcomes students into her home, onto her neighboring farms, and into the working lives of her favorite cast of butchers, bakers, and Armagnac makers. To find out more about her offerings, visit her Web site: kitchen-at-camont.com.